Diana Walker

About the Author

BONNIE ANGELO is the author of the acclaimed
First Mothers. In more than twenty-five years as
a correspondent for *Time*, she reported on the
White House kaleidoscope in both the East
Wing and West Wing. As a Washington
reporter and bureau chief in London and New
York, she has covered newsmakers and major
events in all fifty states and around the world.
She lives in Bethesda, Maryland, and New York.

ALSO BY BONNIE ANGELO

First Mothers: The Women Who Shaped the Presidents

First Families

*The Impact of the White House
on Their Lives*

Bonnie Angelo

HARPER

NEW YORK • LONDON • TORONTO • SYDNEY

HARPER

A hardcover edition of this book was published in 2005 by William Morrow, an imprint of HarperCollins Publishers.

HarperCollins books may be purchased for educational, business, or sales promotional use. For information please write: Special Markets Department, HarperCollins Publishers, 10 East 53rd Street, New York, NY 10022.

FIRST HARPER PAPERBACK PUBLISHED 2007.

The Library of Congress has catalogued the hardcover edition as follows:

Angelo, Bonnie.
    First families / Bonnie Angelo.— 1st ed.
        p. cm.
    ISBN 0-06-056356-7 (alk. paper)
        1. Presidents—United States—Biography—Anecdotes.
    2. Presidents—United States—Family relationships—
    Anecdotes. 3. Presidents' spouses—United States—
    Biography—Anecdotes. 4. Children of presidents—United
    States—Biography—Anecdotes. 5. White House
    (Washington, D.C.)—Anecdotes. 6. Washington (D.C.)—
    Social life and customs—Anecdotes. I. Title.

E176.1.A665   2005
973'.09'9—dc22
[B]
                                                    2005041474

ISBN: 978-0-06-056358-5 (pbk.)
ISBN-10: 0-06-056358-3 (pbk.)

07 08 09 10 11 WBC/RRD 10 9 8 7 6 5 4 3 2 1

For Kip
good son, fun friend

The White House shaped my character. . . . What an
extraordinary opportunity was mine!
—*Luci Baines Johnson*

She will be hurt when she discovers that whatever she does
and no matter how hard she tries, the new First Lady is going
to be under attack. . . . She will wonder if she can endure it for
four years.
—*Lillian Rogers Park,*
White House seamstress, 1929–1959

I do not think any two people ever got more enjoyment from
the White House than Mother and I did.
—*Theodore Roosevelt*

Contents

⟞⟝

Preface

This book has germinated in a corner of my mind for more years than I care to think about. It started, I suppose, during the Eisenhower administration, on the first day I showed my brand-new White House press pass to the friendly guard at the gate of the most famous home office in the world. Almost not believing where I was, I entered the messy hubbub of the press room in the West Wing and, green as grass, joined the scrum of reporters gathered around Press Secretary Jim Hagerty's desk. I have no idea what the story was that day; I only remember that it was the beginning of my passion for the White House and the families who pass through it on their way to history.

In those waning days of the Eisenhower era, I got to see the world-famous President in his two official roles: at press conferences, where the vein in his right temple throbbed visibly if he was asked a barbed question, and at state dinners, when as a press-pool reporter (very proper in evening gown and long white gloves) I sometimes even chatted with Ike, the genial host, dashing in white tie and tails. And Mamie, with her trendsetting bangs and gift of easy small talk, unfailingly charmed women reporters as well as guests. The last of the First Ladies born in the nineteenth century, Mamie, whose "project" was Ike and White House entertaining, was a natural transition to those who would follow with national causes and public speeches.

The arrival of Jacqueline Kennedy, with her determination to return the White House to its fading glory, further stirred my interest in the President's House and its residents—and the impact they have had on each other. Those who live there love it but sometimes resent it; they use its power to advance their chosen concerns but sometimes feel they are pampered prisoners.

A new First Lady moves into the White House proud of her husband's victory but anxious about the demands it will make on her and what it might do to her family's life. Several nineteenth-century wives wilted under the pressure; most twentieth-century First Ladies blossomed in their new role. Even the uncertain ones are usually won over by its perks and pleasures and in due course are even reluctant to go back to their quotidian lives.

For more than two hundred years this one house, and the people who live there, have reflected the enthusiasms of a nation whose bedrock character has always been upbeat and positive, laced with an unapologetic delight in pleasure. And in return the American people have shared in the joy and grief of the families who—often to their private distress—belong to the country for a few years. They have showered White House brides with presents they will never use, they wrote 300,000 letters to the Clintons' dog and cat, and when young Frances Cleveland gave birth to the first child of a President born in the White House, half of the women in America knitted booties.

Years of observing a series of White House families, of covering them on the South Lawn and in remote corners of this country and around the world and getting to know them as individuals, led me back to their earlier counterparts. I wanted to discover more about those who made the President's House a personal home and those who were overwhelmed by its demands. To me, the marvel of the elegant yet malleable Executive Mansion is that it has gracefully adjusted to every family structure from the childless James Polks to the Theodore Roosevelts' rumbustious six, plus the four widowers who assembled a variety of households.

For modern teenage daughters, however, the relationship is tenuous at best, thus I am especially grateful to Lynda Johnson Robb, Luci Baines Johnson Turpin, and Susan Ford Bales for sharing with me their memories and reflections of their White House years, and to Margaret Truman Daniel for giving me the girl's-eye view of the clash of wills that continues today.

Beginning early in the twentieth century, a commendable new attitude has added much to our knowledge of White House life: eleven First Ladies and another dozen daughters, plus a few sons, have written their own accounts of life in the President's House. (What a pity Jackie Kennedy did not turn her wit and erudition to that end.) Julia Grant, in 1887, was actually the first of this unique group to write her memoirs, but her flowery product failed to find a publisher until 1975.

I dug deep into that trove, along with books from White House employees whose combined service adds up to more than 250 years and whose recollections and insights offer no-holds-barred candor. From within the First Ladies' East Wing offices a number of key professional staff have shared their views of the families offstage.

Among the more enlightening books by outsiders are rare volumes by the "lady journalists" of the nineteenth century, who provide an almost video-clear picture of the families and entertaining in the President's House, including gossip and social contretemps that set the capital atwitter.

In general, for the average American, even those who are (like me) political junkies, White House life in the mid-nineteenth century is a black hole in our study of history. That largely forgotten period came alive for me thanks to the preeminent journalist of the "National Metropolis," an engaging writer who covered seventeen presidents—John Quincy Adams to Grover Cleveland—over a span of sixty years, from 1827 to 1887. Even his name is fascinating—Ben: Perley Poore. In his spellbinding two-volume tome, titled *Perley's Reminiscences of Sixty Years in the National Metropolis* he never explained the whimsical colon.

Poore observed everything—from violence erupting on the Senate floor, backroom politics, and Lincoln's problems with his Civil War generals to the decorations and gargantuan menus of White House dinners, and every flounce and scallop of a First Lady's gown—and delivered all with the spice of a theater critic. In an era when First Ladies were rarely noted by serious journalists, except when they caused trouble, Poore held an advanced view: "It is well that the memories of the gentler sex, who have from time to time taken prominent part in shaping the destinies of the nation, should also be remembered." He saw to it that they are.

Over the years of observing First Families, traveling with First Ladies, interviewing them, and writing about them, I have come to feel great admiration for their willingness to use their luster and energy to benefit good causes and sympathy for the intrusions and stresses they must bear. My hope is that this book will encourage readers to share my appreciation of thirty-nine special women in our history and the children for whom the President's House was, for a while, a most unusual home.

Stepping into History

WASHINGTON SPARKLED LIKE an icicle after a blizzard roared in as the 1961 inaugural celebrations were beginning. On any other day a snowstorm would have paralyzed the city, but on this day it was an opportunity to show the resilience, fortitude, and jaunty spirit of the New Frontier. At the White House the near-endless inaugural parade straggled by the presidential reviewing stand, the trumpeters in the marching bands struggling to make their fingers and the valves function in the brutal cold—while trying, like the thousands who lined Pennsylvania Avenue, to catch a glimpse of the new President and his wife.

Never in its long history had the White House welcomed a presidential couple as storybook young and glamorous as John Fitzgerald and Jacqueline Kennedy. He was forty-three, the youngest elected President; she was thirty-one, mother of an infant son and three-year-old daughter. Together they made a family who would beguile the nation. The new President, whose mop of sandy hair was the cartoonists' delight, wore—or at least carried—a top hat, in deference to tradition and the hatmakers union; the new First Lady instantly set a new threshold of chic, from her pillbox hat—copies would be in department stores within days—to her old-fashioned fur muff and fur-trimmed ankle boots.

At the inaugural ball that evening Jackie was a generation apart from Mamie Eisenhower's full-skirted, sequin-studded gowns. Jacqueline Bouvier Kennedy was sophisticated elegance in a stalk of white silk softened by a cape of fluttering chiffon. She owned the night; it was her debut as a megastar, and she infused the White House with a magic she would never lose.

Theirs was the ultimate changeover at the Executive Mansion. In the finest presidential tradition, the transition was courteous and circumspect, even though Kennedy knew that Ike considered him a young whippersnapper and Eisenhower was aware that Kennedy viewed him as a symbol of the past. As the youngest couple arrived, the then-oldest President and his wife, the much admired Dwight and Mamie Eisenhower, were departing after eight happy, conventional years. They were an aging army couple retiring to their Gettysburg farm, giving way to a growing family that required a nursery.

The Eisenhowers represented success born of bedrock middle-American beginnings, the comfort of the familiar, a nation basking in good times. The Kennedys brought a sweeping reordering to the White House—a new era and a new generation; a change of political party, of style, of goals; the dashing war hero replacing the legendary general who had masterminded victory. In a matter of hours the White House would have to adapt to those differences.

Both the incoming and departing First Ladies felt some reluctance on Inauguration Day. Mamie, lighthearted, sociable, a popular favorite, was sorry to leave the home she and Ike had lived in longer than any of the thirty in their thirty-seven years of marriage, the beautiful house she had made her own, with its well-trained staff to do her bidding and a social life she had enjoyed to the fullest. She had made it a second home for her grandchildren, who lived close by, and the regular meeting place for her canasta group. From her first day as First Lady, Mamie, the five-star general's wife, was not intimidated by the White House.

Jackie entered her new life in the White House with a sense of dread, a fear that she would be caged, that her children would be harassed, their childhood spoiled. With her elite pursuits and standoffish manner, she had been targeted as a campaign liability. Already she had learned that she could not even control her own name: she wished to be "Jacqueline" outside her select circle of family and friends, but she had become "Jackie" to all. She disliked "First Lady" as a title—"It sounds like a saddle horse," she protested—but nonetheless she was tagged "First Lady." Later she reflected, "I felt as if I had just turned into a piece of public property."

That was only a slight exaggeration. From the moment a new First Lady crosses the North Portico and enters the White House on Inauguration Day she becomes a public figure, whether she likes it or not. The White House is at the juncture of its family's personal lives, their private joys and sorrows, and the life of the nation they represent; family moments and historic events are intertwined.

In addition to managing her children and the inescapable day-to-day planning, the First Lady is required to be chatelaine of the world-famous mansion and is expected to appear supportive of her husband—or at least not damaging—as he wrestles with the world's most powerful position. The White House is a home, a museum, an institution, a symbol—and for the families who live there, it is both a palace and a prison.

Even Hillary Rodham Clinton, a lawyer long involved with public issues and for twelve years a governor's wife, was taken aback by all her new position entailed. "I don't think anyone is prepared for the whole role that comes with being First Lady," she said in retrospect. "It is not a 'job'—it is an intense, overwhelming experience. There is no guidebook to tell you what to do."

Literally overnight, the new First Lady wakes up to find herself a different person, one she may not recognize or wish to be. Without her consent she is transformed from private helpmeet into a front-page figure. She quickly learns that in return for its many perks and four-star services, the White House makes its own demands on its residents; while it enhances their status it curtails their lives and imposes unwanted duties. Even before the inauguration, Jacqueline Kennedy had received invitations to attend almost three hundred events and more than a thousand requests to lend her name to every kind of organization. She declined almost all, which, her staff director, Letitia Baldrige, well remembers, led to anger and pressure.

The title "First Lady" exposes its bearer to both praise and criticism, makes her more cautious, less spontaneous, and the White House spotlight makes it all but mandatory for her to find the self she wants to present to the public and to adopt a civic mission.

It can be daunting. Even Eleanor Roosevelt, who would possess the White House longer and more completely than any other First Lady, had qualms. After her husband's election in 1932, she confided to a friend, "I never wanted to be a President's wife, and I don't want it now." In her memoirs, Eleanor recalled her early days as wife of the assistant secretary of the navy: "I used to drive by the White House and think how marvelous it must be to live there. Now, I was about to go there to live, and I felt it was anything but marvelous." But once there, she filled it with family, friends, activists, thinkers, performers—and harnessed its power to advance her interests. She blazed a trail for those who followed.

That activist trail terrified the First Lady who followed Eleanor. Bess Truman, returning from Franklin Roosevelt's burial at his lifelong home in Hyde Park, was in a panic. She had been the Vice President's wife for only twelve weeks, living her same quiet life in a modest apartment when a stroke felled Roosevelt. Bess, pouring out her fears to her old friend Frances Perkins, the secretary of labor, cried, "I don't know what I'm going to do. I'm not used to this awful public life." Eventually she did what she had to but never liked it.

In 1937 the wide-eyed wife of the new congressman from Texas stood outside the iron gates of the White House and snapped a photograph. More than a quarter of a century later, Lady Bird Johnson (born Claudia Taylor) recalled the moment in her diary: "I never imagined that some day I would live on the other side of the fence. Yet even then I felt that this property belonged to me, as it does to every American." When that "some day" came on December 7, 1963, two turbulent weeks after President Kennedy's assassination, she said, "I feel as if I am suddenly onstage for a part I never rehearsed." She mastered the role—her leadership as an early environmentalist (she never liked the term "beautification") would alone qualify her for a First Ladies Hall of Fame.

Shortly after moving into the White House, the articulate Lady Bird shared her first impressions with a group of newswomen: "It's hard to feel at home in a house that belongs to 180 million people. I sometimes hear history thundering down the corridors. . . . For about two weeks I sort of tiptoed and whispered—but now the day's work stretches out in front of me each morning and I don't tiptoe anymore." She looked forward to her White House residency as "a short-term lease on an exciting experience. I want to plumb to the depth its stories, its beauties, its traditions."

With University of Texas degrees in both history and journalism, Lady Bird became one of the Executive Mansion's most appreciative occupants. "The President and his family," she observed, "cannot forget that they have joined the stream of history, that the home they occupy briefly, living their own family life, is also an American showplace where the public is welcomed." Sadly, that traditional open door—unique among the residences of heads of state—is now sharply limited by the threat of terrorism.

Laura Bush, who had been in the White House frequently during her father-in-law's presidency, shares Lady Bird's awe of its history. "Just the fact of living in a house that Abraham Lincoln once lived in is unbelievable," she says. "It's really, really fabulous. And the whole history of the country, I think, is documented by the lives of the people who live in the White House."

Betty Ford, whose route to the White House was unique in history, remembers that when her husband was sworn in as Vice President, after Spiro Agnew had been forced to resign in disgrace, President Nixon jokingly said to her, "I don't know whether to offer congratulations or condolences." Then came Watergate and Nixon's own resignation—and the President escaping impeachment was replaced by an unelected Vice President who had replaced a Vice President facing indictment. Amid that turmoil, Betty suddenly had to face all that comes with the President's House—"I felt like I'd been thrown into a river without knowing how to swim." She dived off the high board before they could pack up their Virginia house and move into the Executive Mansion; she found herself the hostess at an already scheduled state dinner for King Hussein of Jordan, a staunch ally of America who had spent more time in the White House than she had.

Betty was a First Lady who had never been circumscribed by a national campaign. As a result, her spontaneity and breezy confidence to say what she thought had not been sanded down by presidential politics, which made her a lively—and controversial—First Lady. Thirty years later, fashionable at eighty-six and still outspoken, she opposed Republican Party policy on national television, staunchly supporting the Supreme Court's divisive decision (*Roe* v. *Wade*) upholding a woman's right to choose abortion. "It was a great, great decision," she asserted. "I'm worried that it might not continue. The choice should be a woman's—it shouldn't be left to legislators." The GOP might fume, but Betty Ford, as First Lady and now, puts conviction above politics.

While many First Ladies enter the White House with trepidation, others are eager to seize the opportunities it offers. A confident Dolley Madison arrived as a social lion with eight years of experience as Jefferson's frequent hostess; Rosalynn Carter brought an active agenda in the field of mental illness begun in her years in the Georgia governor's mansion; Hillary Clinton did not conceal her eagerness to be a partner in policy making.

Yet Nancy Reagan, after eight highly visible years in the California governor's residence, was awed by the difference between the two mansions: "It didn't really hit me until I walked into the White House as First Lady." She had felt confident about her new role. "I'd been First Lady of California for eight years," she said on the *Today* show twenty-three years later. "I thought, 'I've seen it, I've done it— it can't get any worse.' But it did." Throughout two terms in the White House, she endured a drumbeat of criticism.

Surely no First Lady anticipated the role as far in advance as Barbara Bush. In an oral history, her husband's aunt Mary Walker recalled a moment, long before her nephew was a political name, when the women of the family were idly musing about how anyone could like being First Lady. Barbara, the young wife, spoke up quite seriously: "I'd like it, because, you know, I'm going to be First Lady sometime." And when her amazing prophecy proved true, she welcomed the role, performed it with zest, and became a lasting national favorite.

NINE HOURS INTO his presidency, an ebullient Ronald Reagan waited with his family in the holding room before the first of a string of inaugural balls. His son Michael later pictured the scene: "Dad is looking in the mirror, straightening his white tie, and then he cocks his head and gets that little twinkle in his eye . . . and turns around, and jumps straight up in the air, clicks his heels together, and says, 'I'm the President of the United States of America!'" It was the role the former movie actor dreamed of—he would be the star of the planet—but over his eight years in office he would find that the White House can be a demanding director.

Not every President has been so exuberant as he approached his new life in the White House. Its first full-term resident, Thomas Jefferson, took a jaundiced view of the still-unfinished President's House when he moved in on March 4, 1801, calling it "a great stone house big enough for two emperors, one Pope and the Grand Lama in the bargain." (Charles, Prince of Wales, visiting in 1970, commented, "It really is a little house," so it all depends on what one is used to.)

Surely Jefferson spoke in jest—after all, as an accomplished architect he had been deeply involved with the planning of the President's House from the earliest stages. Perhaps there was a touch of pique, since he had entered the competition for the design, using a false identification, "A.Z.," only to have his plan rejected. (Later, as a founder of the University of Virginia, Jefferson recycled his vision with its distinguishing columns and dome for the university's signature building.)

As decade follows decade, as great moments pile one upon another, the White House has assumed the dimension of a shrine. Anyone entering the fabled mansion for the first time feels it: a sense of awe as the tangible presence of history engulfs you; you want not to speak but to absorb. President Jimmy Carter recalled, "To go into those historic halls was an overwhelming experience for me. I was immersed in a sense of history and responsibility." George H. W. Bush was moved by the same deep sense of respect for "that majestic office." President Ford, who found himself in the White House without really trying, joked, "It's the best public housing I've ever seen." He understood that his was a crucial mission: he had to heal the sundered nation after Richard Nixon resigned rather than face impeachment. And he did.

Most new Presidents would admit to some concern about living there. Bill Clinton, amid the jubilation of winning, commented to Paul Begala, a close adviser, that the White House had been called "the crown jewel of the federal penal system," and said he was determined not to let it imprison him. He succeeded in that better than most.

Two months after he was thrust into the presidency, Harry Truman, a history lover since boyhood, was alone in his White House study, writing to Bess, who was spending an unconscionably long summer in their old home in Independence, Missouri: "I sit here in this old house and work . . . all the while listening to the ghosts walk up and down the hallway and even right in here in the study. The floors pop and the drapes move back and forth—I can just imagine old Andy and Teddy having an argument over Franklin. Or James Buchanan and Franklin Pierce deciding which was the more useless to the country. And when Millard Fillmore and Chester Arthur join in for place and show, the din is almost unbearable."

A couple of years later, while Bess was yet again in Independence, Truman continued to converse with his diary: "This great white jail is a hell of a place in which to be alone." Rattling around in the creaking mansion, Truman again conjured up the president-ghosts who "moan about what they should have done and didn't . . . the tortured souls misrepresented in history are the ones who come back." Still, the "great white jail" proved to be not too hellish to deter him from running for a full term in 1948 and—to universal astonishment—winning four more years with his ghostly colleagues. He had called it both a "glamorous prison" and the "great white sepulcher of ambitions"—yet his daughter, Margaret, recalled that when a band in his inaugural parade broke into "I'm Just Wild About Harry," the President, jaunty in top hat and striped trousers, "danced a little jig" in elation.

Like Truman, Theodore Roosevelt's imagination was stirred in the great house, though his centered only on Abraham Lincoln. "I think of Lincoln, shambling, homely, with his strong, sad, deeply furrowed face all the time," he wrote to a friend. "I see him in the different rooms and in the halls. He is to me infinitely the most real of the dead Presidents." A President and his family live with history; they walk in the footsteps of the great and the failures. All are changed, deepened, made more understanding, by the experience.

Until the new occupants get used to their new residence, the ghost-Presidents may roam its halls more freely than they do. It can be hard to find your way around the mansion —it is difficult even to determine how many rooms there are. Depending on which halls, nooks, and crannies are counted, the family's personal living quarters stretch, by some counts, to as many as fifty, but the White House curator's complete inventory from sub-basement to attic, including every service area, storage space, bomb shelter, mansion office, mechanical equipment, and maintenance workroom comes to precisely 132 rooms and thirty-five baths.

President Kennedy, playing host to the prime minister of Denmark at his first state luncheon, greeted the officials in the private quarters, then escorted them downstairs to the Blue Room. Stepping out of the elevator, he marched them straight into a pantry. Chief Usher J. B. West (White House manager) remembered that "with his usual aplomb, he laughed, 'Oh, this is another room I wanted to show you.'" The amiable Danes were much amused by their special tour, which made a good story back in Copenhagen.

WHO WAS THIS fumbling his way in the dark after the last guest at the inaugural reception of 1853 had gone home? The lights had been snuffed; the White House servants had left for the night. But there was the new President, Franklin Pierce, and his private secretary, Sidney Webster, groping their way by the light of a single candle to the unknown territory of the second floor, searching for bedrooms. And when found, the rooms were in disarray, the beds still rumpled after the Fillmores' departure.

The dark and unwelcoming White House was a metaphor for Pierce's life on that day, which should have been a time of celebration. He had arrived in Washington alone, cloaked in grief; his wife, Jane, had come as far as Baltimore, but there she stayed for several weeks, unable to face life in the White House. Just two months earlier the Pierces and their eleven-year-old son, Benjamin, had been in a devastating train accident. Jane and Franklin were not injured, but little Benny was mangled in the wreckage, killed as his parents watched in horror. He was their third son, their last child, and the third to die. The first had lived a fleeting three days; the second had died of typhus when he was four. How could any woman, emotionally demolished, bear up under that?

Jane, deeply withdrawn and religious in the extreme, loathed politics and Washington, even insisting that her husband resign from the U.S. Senate a year before his term was up. To please her, he turned down appointments and refused to run for governor of New Hampshire. Then on a June day in 1852 a messenger galloped up to their carriage with news: on the forty-ninth ballot the deadlocked Democratic convention had come up with Franklin Pierce as its compromise candidate. Franklin was ecstatic; Jane fainted dead away.

She taught Benny to hope that his father would not be elected; after the accident she decided, in her near-deranged state, that God had taken her last child in retribution for Franklin's return to the evils of politics. The very thought of the White House shattered this unwilling First Lady. Eventually she took her place at the requisite state dinners and receptions, but her sensitive face would always be etched in deepest melancholy.

If the Pierces were the saddest new residents, Lucy and Rutherford Hayes must have been the gladdest. For four agonizing months after the disputed election of 1876, they had not known who would be declared the winner. At last, only two days before the inauguration, the outcome was decided: at 4 A.M. on March 2 a special electoral commission with a one-vote Republican margin balloted along party lines to declare Governor Rutherford B. Hayes of Ohio the nation's nineteenth President—though the Democratic governor of New York, Samuel J. Tilden, had won the popular vote by almost 265,000 votes, a huge margin in those days.

This most contentious election in American history was concluded amid charges of skullduggery and extremist threats of inaugurating Tilden by force. In that vitriolic atmosphere, complicated by the fact that Inauguration Day fell on Sunday, an unprecedented precaution was taken: on Saturday night, prior to a farewell dinner given by President and Mrs. Grant, Hayes was secretly sworn in as President at the White House.

A public ceremony at the Capitol followed on Monday, but since it had not been possible to organize a triumphal parade and festive ball to honor To Whom It May Concern, Hayes supporters could only welcome him to the White House with a last-minute torchlight parade. To this day the disputed election remains a fractious subject among political scientists. There is widespread agreement with the view of historian Roy Morris, Jr., who dissected the tangled tale of missing ballots, nocturnal lobbying, and deals cut, and flatly concluded that "the stolen election" rightfully belonged to Tilden. But in 1877 Lucy and Rutherford Hayes claimed the White House in the happy certainty that the prize was theirs. A confident Lucy hung new draperies and without hesitation set about emplacing the Hayes agenda in the White House, starting with her total, and much derided, ban on alcohol.

Since then, the only similarly belated decision—the disputed Florida vote in 2000—was decided more quickly, by the Supreme Court, also by a one-vote margin.

CONCEIVING A NEW capital city from scratch was a bold idea, one that even today would be an awesome leap of confidence. In 1800 the important thing was that this infant enterprise, swaddled in great aspirations, should grow into a capital of great power and influence. For a century the occupants of the President's House arrived with a special burden since the single structure, housing both family home and executive offices, was at the intersection of ponderous affairs of state and purposeful conviviality. It fell to Abigail and John Adams to create the template for all who would follow.

On November 1, 1800, in the scraggly village newly named Washington, a place of mud and mosquitoes built on nothing but rough fields, a majestic river, and a vision, history turned a page. This unlikely setting could now boast that it was the seat of the fledgling American government, the great experiment in rule by the people. With all the optimism of the country it would symbolize, the President's House opened its doors to receive its first resident, President John Adams, the correct, even haughty, New Englander who had fought this shift of power from Philadelphia to the South.

It was not a sanguine occasion for the second President. Within days the House of Representatives annouced that, after thirty-six ballots in which Thomas Jefferson and Aaron Burr were tied, with Adams in humiliating third place, Jefferson won the presidency. Adams, as his carriage jolted south, realistically faced the ignominy of rejection for a second term, defeated by his Vice President, his onetime friend and political opposite. The proud Founding Father knew this would be the end of his lustrous career.

Nor was the new residence a welcoming sight to the weary President after the long journey, accompanied by a single manservant, an unimaginable contrast to the entourage that today scurries around a President on his every move. The house itself was an austere rectangle of pale sandstone, stern as a place of detention, not yet softened by as much as a shrub, lacking the welcoming porticoes that would be added decades later. The grounds were a hodgepodge of workmen's shacks, heaps of supplies, and rubbish. President Adams' first thought was to seek other quarters.

Looking around the unfinished mansion—then the largest residence in America—Adams tried to see the outlines of grandeur, but the reality facing him would tax any imagination. It was a glove with no hand inside. He and his esteemed Abigail would strive to put on a good face, to accept the random cluster of buildings as the Federal City—the new American capital, no less—and commit themselves to the social duties required of the President and his lady. All of this had to be carried out with grace in a dwelling that was more barn than mansion.

Two weeks later, Abigail, traveling with her four-year-old granddaughter, Susanna, and seven others, joined her husband. The First

Lady's arrival was less than auspicious. On the last leg of the journey, the thirty-six miles from Baltimore to Washington, the driver of her coach lost the way, struggling through eight miles of dense forest and skirting fallen trees and marshy sinkholes. "We were wandering for more than two hours without finding a guide, or the path," an irritated Abigail wrote to her sister. "The woods became impossible. It was clear we were lost."

At length she was rescued by a Maryland landowner, Colonel Thomas Snowden, who invited the President's wife and her party to stay the night at his home. All in all, it was a miserable journey to a place she had strongly opposed, to an unlivable house, to join her husband in their defeat. Compounding her desolation was the knowledge that their second son, Charles, was dying of alcoholism in New York.

In letters to her sister and her daughter, "Nabby," she poured out her woes: "Shiver, shiver. Surrounded by forests, can you believe that wood is not to be had, because people cannot be found to cut and cart it! The house is made habitable, but there is not a single apartment finished. . . . We have not the least fence, yard, or other convenience, without, and the great unfinished audience room [now the stately East Room] I make a drying-room of, to hang up the clothes in. The principal stairs are not up, and will not be this winter."

Much of her furniture had arrived heavily damaged, and half of her prized china tea service had not survived the rough journey from Philadelphia. Her mirrors were too small for the high ceilings, and there was not "a twentieth part enough lamps." Adding to her dismay, the ship bearing her clothes had not arrived. But being a true patriot, Abigail fulfilled her role, emulating her idol, Martha Washington.

With studied neutrality about "this great castle," she commented, "This House is built for the ages to come" and observed that "it is capable of improvement." And, yes, she conceded, the house was built "upon a grand and superb scale." But of her complaints, she cautioned Nabby, "You must keep all this to yourself and say the situation is beautiful, which is true."

Abigail would live in the President's House for three months, long enough to experience all of its early discomforts and none of its later pleasures. It fell to her to bring its grand but empty rooms to life, to turn the great house into a home—for Mr. Jefferson! It was indeed a bitter task. As soon as their social season was completed, Abigail, her rheumatism aggravated by the cold, damp Executive Mansion, her heart heavy with the defeat, hastened back to Massachusetts and their own snug and reassuring home. John dutifully stayed on until the inauguration, which, pointedly, he did not attend.

MARTHA WASHINGTON, FIRST of that unique handful of Presidents' wives (thirty-nine in all, with two Presidents accounting for four), looked upon her trailblazing role with equanimity. "I know too much of the vanity of human affairs to expect felicity from the splendid scenes of public life," she wrote to a friend. "I am still determined to be cheerful and to be happy in what ever situation I may be; for I have learned from experience that the greater part of our happiness or misery depends upon our dispositions, and not upon our circumstances."

The first First Lady's philosophy should be embroidered on a sampler for those who follow in her footsteps. It served her well. Beloved by the public, she established the role of the President's wife as partner, the new democracy's version of a royal consort, a role that has been redefined again and again to adapt to individuals as different as Ida McKinley and Hillary Clinton, to reflect times as formal as Martha's and as casual as Laura Bush's. And as the individuals adapt to the White House, the White House adjusts to them.

Some First Ladies have been reluctant—even afraid—to leave their own homes, their comfortable circle, their familiar towns. In

1870 Laura Carter Holloway, a Washington observer who wrote one of the earliest books about the Presidents' wives, *Ladies of the White House,* noted with approval that Mrs. John Quincy Adams "gave up the comforts of her home and took possession of the White House." It hardly seems a sacrifice to move the few blocks from F Street into America's premier residence, but that is the way many have felt.

A contemporary First Lady understands that despite the negatives that come with life in the Executive Mansion, it gives her a forum to advance worthwhile causes in her own interests. Jackie Kennedy arrived with her plan to create a more authentic White House; Eleanor Roosevelt brought a daring social agenda; Lady Bird Johnson turned her spotlight on the environment; Rosalynn Carter focused on mental health problems; Hillary Clinton hoped to continue her longtime work for children's issues and health care; both Barbara and Laura Bush have seen it as a platform for promoting literacy and encouraging children to read. Technically, a First Lady, having no job description, can do as much or as little as she chooses, but in this era doing little is realistically not acceptable. "I think people want the First Lady to use whatever expertise they have to help America," said Laura Bush. "I mean, it's a huge platform."

FOR MANY NINETEENTH-century First Ladies thrust into a public life, the prospect was intimidating, even overwhelming. Lucretia Garfield, as the wife of a congressman and senator, knew Washington well enough to dread the White House. She spoke for many First Ladies of her era when she lamented on her husband's victory in 1880, "What a terrible responsibility to come to him—and to me." "Crete," who had been a teacher, had no chance to put her imprint on the White House—her husband's presidency ended with his assassination, lingering illness, and death after just 199 days.

Two Presidents' wives never reached the White House at all. Andrew Jackson's Rachel saw her hero husband win the prize but died ten weeks before his inauguration, not yet bearing the title of First Lady. President William Henry Harrison's wife, Anna, planned to join her husband in the White House when she recovered from an illness, but thirty-one days after his inauguration the old general died of pneumonia. Anna never finished packing.

At the other extreme were such ambitious wives as Mary Todd Lincoln, Helen Taft, and Florence Harding, who had devoted themselves—some would say schemed—to advance their husbands' presidential aspirations. Helen "Nellie" Taft, eager to assume the First Lady role, wrote in her memoirs, "I could not help but feel something as Cinderella must have felt when her mice footmen bowed her into her coach and four. I stood for a moment over the great brass seal bearing the national coat of arms, which is sunk in the floor in the entrance hall. 'The Seal of the President of the United States,' I read around the border and now—that meant my husband!"

Nellie virtually leapt over the great seal. A year earlier Theodore Roosevelt, Taft's then-champion, had offered him the choice of a Supreme Court seat or the presidential nomination. "The court," replied Taft, delighted. "The presidency," said Nellie firmly. In a private letter to Roosevelt, Taft pled, "I would much prefer to go on the Supreme Bench for life than to run for the Presidency. In twenty years of judicial service I could make myself more useful to the country than as President." He was right, but Nellie prevailed. Thus Taft spent four years in a job he did not like and conducted poorly, while his wife basked in her White House glory and influence.

The pushing and pulling that come with a First Lady's role is best understood by the Executive Mansion staff who are around her every day. White House seamstress and personal maid Lillian Parks, drawing on thirty years of serving four very different women, predicted what every new First Lady would feel: "It will take months to get used to the mausoleum feeling of the rooms, which belong to her and yet don't belong to her . . . she will be terrified at those first

State Dinners . . . never will she be completely free of the fear of danger to her children and her husband . . . she will feel hurt when she discovers that whatever she does, and no matter how hard she tries, the new First Lady is going to be under attack. Somewhere along the line, the new First Lady will wonder aloud whom the public elected—the President or her. She will be shocked to find out how much is expected of her. She will wonder if she can endure it for four years—and then maybe another four." The six living former First Ladies, and one current, would likely agree that the observant seamstress got it right.

"Out They Go, In We Go!"

⟶⟫·⟪⟵

I T WAS THE sweetest of triumphs, that Inauguration Day in 1893. Never before—and never since—had a President recaptured the White House after voters banished him to four years in the political wilderness. Grover Cleveland's partisans cheered his comeback with a lusty song: "Grover, Grover! Four more years of Grover! Out they go, in we go; then we'll be in clover."

On that singular day both the incoming First Family and the White House staff were fully prepared for one another, with mutual welcome for their return. But more often than not the changeover involves varying degrees of shock.

There was, for example, the arrival of the Lyndon Johnsons in 1963 with their two daughters, the first teenagers in the White House for fifty years. Luci, an effervescent sixteen-year-old, made an immediate impression: she and a friend wanted a cozy fire in her new bedroom's fireplace—and what they got was a room filled with billowing smoke. She remembers standing on a desk in her nightgown, trying to open a window—"and there was a White House policeman looking straight at me." The two girls, forming a little bucket brigade with water glasses, were frantically trying to douse the blaze when White House attendants rushed to the rescue, opened the damper, and all was well. It was a fine introduction to the mansion's new teenagers and for Luci a memorable beginning in her new home—"I spent my first week in the White House getting rid of the smoke on my newly painted walls." Forty years later, she can laugh about the fiasco: "It was my first night in the White House, and I was going to burn the place down!"

Luci's little fire was nothing compared with the conflagration that engulfed the White House in the Nixon years that followed, the events that will always be known as Watergate. (It will be lost in future mists that the two-year episode began when Nixon White House operatives broke into the Democratic National Committee headquarters located in the Watergate Building.) In that turbulent time, the happenstance First Lady, Betty Ford, was certain of one thing: she would not change. "I've spent too many years as me," she declared. "I can't suddenly turn into a princess." But without missing a beat, the former Martha Graham dancer fell into step with the White House pattern and soon jazzed up the tempo.

Even in a normal transition, the overnight change can be disorienting, and for the White House domestic staff the change from Bess Truman to Mamie Eisenhower was just that. Plainspoken, plain-style Bess, whose egalitarian attitude made her a staff favorite, was dutifully fulfilling the First Lady role she did not want; Mamie, the sociable army wife accustomed to moving from post to post, always to better quarters as her husband leapt up the military ladder, took command of the White House with delight.

Each morning she summoned her lieutenants to her "office"—her pink-covered bed. Propped up against a cluster of pink pillows, wearing a pink bed jacket and a pink satin bow in her hair, Mamie worked from notes on her bed tray. Her passion for pink soon spread through the mansion. "Pink and fluffy were what we became in the White House," seamstress Lillian Parks recalled with a touch of scorn. "Pink crept onto the dining room table—even pink food."

But for all her sparkling ingénue charm and girlish delight in over-the-top decorations ("feminine to the point of frivolity" was the slightly censorious way Chief Usher West put it), Mamie was a white-gloved inspector general in *her* White House, demanding spit-and-polish housekeeping and smart precision for White House social occasions. But she was also quick to praise. "She could give orders, staccato crisp, detailed, and final, as if it were she who had been a five-star general," West remembered. "She established her White House command immediately." When West asked the President to okay the menu for a stag luncheon, she snapped, "*I* run everything in my house." Mamie knew what she wanted, and she wanted it done now! And she did it all without denting her image as the sparkling, lighthearted wife of the hero. Lady Bird Johnson, at the other extreme, would tell West, "I want *you* to run the White House—I want to devote my time to other things."

AS CLOCKS SOUNDED midnight, signaling the start of the 1857 Inauguration Day, a band at James Buchanan's doorstep struck up "Hail to the Chief," the stirring tune that is music to the ears of every President. The tradition had begun in 1845 when the very new, very young First Lady Julia Tyler had the Marine Band play it to signal her husband's arrival at the White House ball welcoming Texas into the Union. The effect was such that "Hail to the Chief" (borrowed from a London musical) became a part of the panoply announcing the presence of the chief executive.

For Buchanan, the fanfare sounded triumph on his fourth try for the White House. Following the cheerless four years of Jane Pierce as First Lady, lifelong bachelor Buchanan—the nation's only never-married President—brought the White House alive with his charming, London-polished niece, Harriet Lane, as his hostess. Beginning at his inaugural ball, where the dancing continued until dawn, the Buchanan White House was a glittering social mecca, but by the end of his single term it was a political quagmire.

Three Presidents later, General Ulysses Grant, elected in a landslide after the Civil War Buchanan had done nothing to prevent, entered the White House with an act more significant than trumpet trills: on his first night he dismissed the military guard that had protected the President's House since the guns of war had first blazed. With that order at his own front gate, the celebrated general who had won the war—now commander in chief—signaled that the danger that had for so long threatened the city was finally behind them. The Grants then settled into the President's House for eight contented years—which also ended in a political quagmire. Ironically, the general who could command an army with honor was overrun by appointees and friends using their connection to him as a springboard for corrupt financial schemes, which would always tarnish Grant's presidency.

TYPICALLY, AS SOON as a new President and First Lady unpack in the White House they have yearned for, they want to make changes. The White House, of course, does their bidding without delay—unless it finds ways not to. Changes can be significant, controversial, transient, or frivolous.

Chief Usher West's first day with the Lyndon Johnsons began with an encounter with the President, who was in a state about his shower. Not enough pressure, he growled at West. "If you can't get it fixed, I'll just move back to The Elms [his Washington home]." A ridiculous threat, but the White House manager does not argue with an enraged LBJ who wanted his power shower.

A factory duplicate of The Elms model wasn't right. Another and another and a fifth from the factory were not right. Walls were torn out, a special water tank was installed, and, West recorded, "thousands and thousands of dollars, not counting man-hours," were spent to construct a shower to please Lyndon Johnson—who at the time was ordering staff and family to switch off lights to save money. When the LBJs moved out and the Nixons moved in, the new President looked at the shower, five years in the perfecting, and directed, "Get rid of this stuff," and the new First Lady ordered the White House bathed in floodlights, a thrill for tourists and Washingtonians alike.

Whatever a President wants, the White House delivers—no questions, only action, a "Yes, sir!" service that feeds latent imperial tendencies, not unknown among Presidents (and a few First Ladies). Where LBJ's obsession concerned water, President Nixon's concerned fire—he wanted logs burning in the fireplace in the Lincoln study, his favorite room, regardless of the weather outside. Thus, on a sultry summer evening a houseman would build a cheery blaze while turning the air conditioning ever lower. It was what the President wanted—no questions, only action.

The staff gets used to one set of whims and then a new set arrives. George H. W. Bush banished broccoli; Bill Clinton brought it back. Lucy Hayes banned billiards; James Garfield loved the game. Herbert Hoover banned dancing; Eleanor Roosevelt found it great fun. The Grants served six wines with dinner; Lucy and Rutherford Hayes served none at all. Mamie Eisenhower loved soap operas; Jackie Kennedy preferred grand opera.

When the Trumans moved into the White House, Head Butler Alonzo Fields was humiliated that he couldn't mix an old-fashioned to please the President and his wife—until, on the third try, he simply poured bourbon over ice. Bess sampled, beamed, and declared, "Now, that's the way we like our old-fashioneds!" From time to time the staff is put to the test of true sangfroid: Chief Usher Howell G. Crim, a gentleman of the old school with many years at the White House, never quite recovered from a morning conference with Eleanor Roosevelt, who padded into his office barefoot and clad in a yellow bathing suit. Not an eyebrow was raised—until she was out of sight.

Adapting to changing whims is not easy, especially when the notions are as fatuous as Nellie Taft's injunction against waiters who wore whiskers of any kind or were bald, or Lou Hoover's edict that all butlers must be the same height—yet she hired six-feet-four-inch Alonzo Fields, who stayed for twenty-one years.

Even religious practice can change overnight. LBJ sampled churches all over town; Nixon initiated the first Sunday worship ser-

vices in the East Room, where Billy Graham and other visiting ministers delivered a message of uplift and the kitchen delivered danish
and fruit cup. This White House innovation was criticized for en-
twining church and state; Jimmy and Rosalynn Carter ended the
practice, faithfully attending a nearby Baptist church. Religion returned to the White House with George W. Bush's morning Bible
study (optional) for his advisers, which also drew disapproval for
mixing church and state.

THERE ARE TIMES when the staff would like to cry out against
changes that permanently affect the mansion they cherish. The saga
of the swimming pool was a prime example. For Franklin Roosevelt,
his legs paralyzed by polio, recreation was confined to swimming,
which was therapeutic; boating, which was invigorating; and
martini-making, which was fun. Most important to him was a new
White House feature—the indoor swimming pool. It became a part
of White House history, but thirty-six years later Nixon tore out the
pool to provide more orderly quarters for the media.

"Of all the changes that have been made in the old mansion
from one administration to another," Chief Usher West confided in
his memoirs,

the one that affected me was the removal of the swimming pool.
The house "belongs" to whoever lives there. But I hate to see history
disappear. It was a gift to President Roosevelt from the schoolchil-
dren of America, who collected millions of dimes [in the depths of
the Depression] to pay for constructing the heated indoor pool,
which that President used every day in his first years of office for
post-polio therapy. I remember President Truman swimming there,
his glasses fogged up, as part of his fitness regimen; the Eisenhowers' grandchildren, coming over on week-ends, splashing around
with the greatest glee; then next a colorful sailing scene, commis-

sioned by Ambassador Joseph Kennedy, that brightened up the walls for the swimming races between President Kennedy and his Cabinet [and occasionally with his Welsh terrier Charlie]; the scores of bathing trunks hanging from the hooks for President Johnson's guests—in all sizes from King Farouk to Mahatma Gandhi. I miss that swimming pool.

So did Jerry Ford, the athletic President who so abruptly replaced Nixon. Happily, private donors provided an open-air pool outside the Oval Office, discreetly away from tourist eyes, bringing back a favorite White House recreation—though not sans swimsuit, which was, so they say, the preference of at least one President, enthusiastic swimmer JFK.

THE OVERNIGHT METAMORPHOSIS of the White House was never sharper than the exit of the Herbert Hoovers and the arrival of the Roosevelts in 1933. "You will find us a noisy family," Eleanor Roosevelt warned the staff as the multiple Roosevelts moved in. Laughing, joking, buzzing with earnest talk, the extended family—incorporating close friends and aides who had been invited to live in the mansion, plus a torrent of visiting relatives and connections—was never solemn. Longtime employee Lillian Parks remembered fondly that "the President's 'happy talk' was contagious. Backstairs and front-stairs."

The change affected every facet of White House life—personalities, style, interests, as well as politics. The Hoovers' frosty formality had made for starchy evenings; even when dining alone in the huge State Dining Room they preferred, they dressed in full evening attire and were silent. The Franklin Roosevelts turned the White House upside down literally overnight. "Within a few days," Head Butler Fields recalled, "this family had the old White House rocking with their gaiety and laughter."

Even before Inauguration Day, it had not been a comfortable transition. For any defeated President, the sixteen-block drive from the White House to the Capitol with the victor who unseated him must seem humiliating and endless. It takes a man of sturdy character to rise above rejection. Contemporary Presidents have swallowed hard and have done that: Ford to Carter, Carter to Reagan, Bush to Clinton—all have mustered the poise to play their part with dignity.

Rarely has the hostility of the loser been as overt as Hoover's. After considerable resistance, he followed White House tradition (wisely abandoned since) and invited the President-elect to tea the day before the swearing-in. During the strained courtesy call, as FDR biographer Arthur M. Schlesinger, Jr., recorded it, "Hoover pointedly disparaged Roosevelt, who, deeply angered, left as quickly as was acceptable. On the drive along the crowded route to the Capitol the following morning, Hoover sat motionless and unheeding. At one point, the President-elect suddenly felt that the two men could not ride on forever like graven images. Turning, he began to smile . . . and wave his top hat. Hoover rode on, his face heavy and expressionless." Within two hours, Franklin Roosevelt and his exuberant family were making the White House their own, as it would be for twelve historic years.

Though that was the most overtly hostile transition, there was precedent for ill will among Presidents defeated for a second term. In 1825 the unbending John Quincy Adams followed the example set by his father in 1801, refusing to attend Andrew Jackson's swearing-in even though he was still in Washington, in a friend's house. He took his usual morning horseback ride, ostentatiously aloof from the goings-on at the Capitol, but could not shut out the thunder of cannon proclaiming the new President—"the people's President"—who was cheered by the largest and unruliest crowds ever witnessed in Washington's twenty-eight years. Exuberant frontiersmen had traveled "as far as five hundred miles to see General Jackson," the great orator Daniel Webster wrote to a friend. "They really seem to think that the country is rescued from some dreadful danger."

Bad blood surged again in 1953, over a trivial bit of protocol. World hero Eisenhower insisted that President Truman should show deference by picking him up at his hotel for the ride to the Capitol. Harry refused; he was just as stubborn as Ike and had custom on his side. Ike was forced to drive to the White House to greet and ride with his political enemy; even then he refused to step out of the limousine to shake hands with Truman on the North Portico steps—all this enmity when neither had run against the other. Truman, however, as a surrogate for Adlai Stevenson during the campaign, had sharply attacked the five-star general and political neophyte, who had never been subjected to such insubordination. Their feud continued for years, but as they aged and mellowed, they eventually reached a truce.

The former Presidents' club is too small, the empathy too great, to allow lasting hostility. The kinship, even among the defeated, overcomes all. The nation likes to see its Presidents rise above their differences, as was demonstrated at the dedication of the Clinton Presidential Library shortly after the election of 2004, where winners and losers stood together in the rain in a show of bipartisan solidarity. The Founding Fathers would have been proud that the system, going into its third century, is still strong.

Jimmy Carter, the most gracious of defeated Presidents, entertained about thirty of the incoming Reagan team at tea in the White House a few days before the 1981 inauguration. Reaganite Charles Z. Wick, the incoming head of the Voice of America, would never forget a little vignette from that day: "I remember looking out over the south driveway and there was Amy Carter, in blue jeans, shoveling some of her stuff in the back of the station wagon. It was a symbol of the transfer of power." A civilized and courteous transfer. Thirteen-year-old Amy, unlike her parents, was happy to be leaving the White House—she was going back home to Plains, Georgia, eager to go fishing with her irrepressible grandmother, "Miss Lillian," and to be herself again.

The instant adjustments from President McKinley to Theodore Roosevelt, following McKinley's assassination in 1901, taxed the White House to its limits. The staid McKinleys, with no living children and always careful about delicate Ida's health (she suffered from epilepsy), were replaced by the ebullient Roosevelt—cowboy, big-game hunter, harbinger of the new century—and his much admired wife, along with their six full-of-beans children and a veritable zoo.

"How that vivid extroverted personality has enlivened the city," observed Countess Marguerite Cassini, daughter of the Russian ambassador, following TR's sudden ascent to the presidency. It was said that "you go to the White House to shake his hand and go home to wring the personality out of your clothes." Countess Cassini contrasted "the melancholy McKinley and this gay showman—that terrific smile, a cannibal smile, that torturous grip which is his handshake. The Roosevelt stamp is already on the nation. He is ten men rolled in one. He never sits when he can stand, he never walks when he can run."

The upheaval shook the White House from rooftop to cellar; overnight the stuffy mansion was permeated with a sense of fun— "the wildest scramble in history," Chief Usher Ike Hoover called it, but the staff handled it with only the occasional wince.

Even when the two Presidents are ostensibly friends, the White House changeover is awkward. Theodore Roosevelt graciously invited the William Howard Tafts to spend the pre-inaugural night at the White House, a nice idea that didn't work. Alice Roosevelt Longworth, TR's daughter, found it "a singularly hushed and cheerless dinner." For the Roosevelts, who had been so happy there, "the inevitable melancholy of saying good-by, of closing the door on great times," hung over the evening, while the Tafts "soft pedaled their natural elation." Nellie Taft pointedly commented in her memoirs, "Neither Mrs. Roosevelt nor I would have suggested such an arrangement."

For families who made the White House a true home, the final night in the mansion they shared with America, and the last good-byes to the staff who served them loyally—old friends, really—is an emotional moment not easily shared with the newcomers, who, deep in some hearts, might seem to be usurpers.

The handover of the White House from the rambunctious Roosevelts to the sedate (except for twelve-year-old Charlie) Tafts should have come as a welcome relief to the staff, but it was like air sputtering out of a balloon.

Major Archibald Butt, the indispensable personal aide to both the Roosevelt and Taft households, tried to explain the changes in chatty letters to his mother and sister: The Tafts "cannot be compared to the Roosevelts, for there is simply nothing in common—their life, their mode of thought, their education, and above all their natures are . . . as widely separated as the East is from the West." At his first dinner with the Tafts, Butt later confided, "I could not get the Roosevelts out of my mind. The conversation was natural, but there was something missing. It was like being at the White House with the President absent . . . I simply missed the marvelous wit and gaiety, the personality of Roosevelt and the sweet charm of his wife."

Taft seemed to share Butt's view. When a friend asked the new President how he liked the job, Taft replied with disarming candor. "I hardly know yet. When I hear someone say 'Mr. President,' I look around expecting to see Roosevelt, and when I read in the headlines that the President and Speaker Cannon have had a conference, my first thought is, 'I wonder what they talked about.'"

Major Butt, striving to be fair to the new First Lady, a woman of little humor and many opinions, commended "her honesty and directness . . . her straightforward method," diplomatically omitting that she was heartily disliked by the staff. "The house is very orderly and quiet," he wrote with a trace of nostalgia. "One never hears any noise at all." He could never have said that about the Roosevelt White House.

Taft's phlegmatic nature—he could doze off at any time—occasionally posed a problem. Major Butt recounted a scene at a funeral that might seem hilarious to anyone other than the President's aide: "In the midst of the services I saw the President fall asleep, and I stood horrified when I heard an incipient snore." As Butt was beginning to panic, the Supreme Court justice seated next to Taft also fell asleep. "I let them remain so, for had either snored loudly I made up my mind to lay it to the Justice." Teddy Roosevelt, in such circumstances, would never have snoozed but might have resented being ignored. "My father," his daughter Alice famously quipped, "wants to be the bride at every wedding and the corpse at every funeral."

In 1912 Major Butt, exhausted after almost twelve harried years at the beck and call of two Presidents and First Ladies, took a long-overdue vacation in Europe. He booked his return on the maiden voyage of a splendid new ship—the *Titanic*. The conscientious Archie Butt surely was helping the women and children reach the few lifeboats; he was not among the survivors.

IN THE NEXT decade, death brought Calvin Coolidge into Warren Harding's White House. It would be hard to find two men of less affinity. Harding, an Ohio newspaper publisher and senator, was the embodiment of "cronyism" in the Executive Mansion. Prohibition? That was for other people. The chief executive enjoyed his private stash in the family quarters, where he and his old pals played poker while the First Lady, the formidable Florence, poured the drinks. A powerful force, Florence—whom Harding called "the Duchess"— often wrote her husband's speeches, attended important meetings, and took credit for masterminding his election. "Well, Warren," she reportedly declared, "I have got you the presidency: what are you going to do with it?"

Halfway through his term, Harding, in deep depression due to charges of corruption wracking his administration and a heart so weak he had to sleep bolstered upright by pillows, set out on an eight-week vacation. To take him far away from a White House engulfed by scandal—the plan was for a train trip to the West Coast, where he would have a light schedule of appearances, followed by some fishing in Alaska and a return via the Panama Canal, with a stopover in Puerto Rico. Two doctors and a trained nurse accompanied him, along with the watchful First Lady. The Vice President, vacationing at his family's old homestead in Vermont, was not advised about the grave health concerns.

In San Francisco, as "the Duchess" read a favorable *Saturday Evening Post* article aloud to him, Harding was felled by a fatal stroke—and Calvin Coolidge was President. It was almost three in the morning when the telegram arrived at the Coolidge family home in the tiny settlement of Plymouth Notch. Coolidge woke his father, a notary public. Both men, aware of the enormity of the moment, put on their Sunday suits, and in the simplest of all inaugurals, father administered the oath of office to son by lamplight in the threadbare dining room, with the new First Lady, his driver, a stenographer, and a couple of reporters as witnesses.

Coolidge was characterized euphemistically as "shy" and "taciturn," but shy people tend to be courteous, which Coolidge distinctly was not. As Vice President, he went to dinners only to eat; as White House host, he could only be called antisocial, deliberately making his guests feel uncomfortable. In an oft-repeated story, a dinner guest told him she had made a bet that she could get him to speak more than two words; he replied, "You lose." He was known to go through an entire official dinner without having a conversation, while First Lady Grace covered his silence with her charm. Coolidge's only social grace was his wife by that name.

In contrast to Harding and his high-flying bonhomie, Coolidge ruled that the majesty of the White House obviated the need for "pomp and splendor," thus he would "maintain as far as possible an attitude of simplicity. There is no need of theatricals." This attitude also saved money—and Coolidge was frugal to the core. Exercising his own lèse-majesté, Coolidge often did not offer guests at receptions so much as a glass of ice water. "It would encourage them to linger," he explained to a butler.

Secret Service agent Edmund W. Starling wrote that Coolidge "divided sandwiches as meticulously as robbers splitting the loot [and] sliced cheese with the precision of an Antwerp diamond cutter faceting a crown jewel." Still, "Silent Cal" was not without a sense of humor, mostly in the form of "boyish" practical jokes, which always harbor a touch of malice. It was surprising that Coolidge liked living in the White House better than Grace did. He sometimes sat in a rocking chair on the North Portico and enjoyed himself as if it were the Vermont farm, and more than once the servants were discomfited when the President wandered the family halls in his nightshirt. He especially liked what was, for his lifestyle, the comfortable White House allowances.

Happily Grace Coolidge, cut from a different cloth, found the mansion "a home rich in tradition, mellow with years, hallowed with memories." At the same time, her role troubled her: "This was I, and yet not I," she wrote later. "This was the wife of the President of the United States and she took precedence over me; my personal likes and dislikes must be subordinated to the consideration of these things required of me."

There was one exception to Cal's stinginess—he loved to buy pretty, bright-colored clothes, even with a touch of flapper fringe, for Grace. She was everything he was not—amiable where he was sour, gracious where he was gruff—and the nation adored her. She was his true trophy wife, a First Lady who never made a wrong move. And *he*, Calvin Coolidge, was the man Grace Goodhue had chosen to marry.

EVERYTHING IS DIFFERENT when a Vice President is thrust into office by the death of the President. Yet even in grief, life in the Executive Mansion must go on seamlessly. Imagine, then, the panic of Vice President Andrew Johnson's family, entering the White House in the wake of the first assassination of an American President, the public murder of Abraham Lincoln, a man already stamped by history, a leader marked for many monuments. Feelings against Johnson ran so high that one minister asked to be released from his obligation to pray for the President of the United States. The trauma of it all was more than Eliza Johnson could bear.

Her husband had served in high political offices for thirty-odd years, but she had never entertained the thought of herself as First Lady. When the distraught Mary Todd Lincoln asked to remain in the White House for two months as she tried to pick up the pieces of her life—a burden she would never be able to manage—Eliza was only too glad to oblige. Those weeks gave her time to confront her own predicament: how could a small-town woman from Tennessee, an invalid, old at fifty-eight, suddenly become mistress of the White House? She was one of many First Ladies for whom the very notion of being cast as a public figure, setting styles and social standards, was overwhelming.

She confided her dismay to Colonel William Crook, the indispensable aide who would serve First Families for forty-two years: "Crook, it's all very well for them who like it, but I don't like this public life at all. I often wish the time would come when we could return to where I feel we best belong." She was well aware that her husband, a pro-Union southern Democrat chosen to demonstrate unity, was hated by Republicans as a usurper; southern Democrats reviled him as a turncoat. Others who might have noticed him at all had been shocked by his rambling, inebriated speech at his swearing-in as Vice President, a performance Lincoln attributed to illness—treated by a tumbler of brandy. "Andy's no drunkard," he declared.

In that sad summer of 1865 Eliza immediately delegated all First Lady duties to their elder daughter, Martha Johnson Patterson, the wife of a senator, who pleaded for understanding: "We are plain people from the mountains of Tennessee, called here for a short time by a national calamity, but we know our position and shall maintain it. . . . I hope you will not expect too much of us." (When Martha asked forbearance for her Tennessee-mountain background, she was a bit disingenuous—in her father's Senate years she attended the prestigious Georgetown Visitation Academy and spent holidays with President and Mrs. Polk in the White House, where she met Washington's leading figures, including the incomparable Dolley Madison.) The anguish of this pleasant family—eleven, including five grandchildren—would deepen as Johnson became the first President to face the humiliation of an impeachment trial.

A WEARY MESSENGER on horseback clattered up to the door of Vice President John Tyler's Williamsburg home as dawn was breaking on April 5, 1841. It was the son of Secretary of State Daniel Webster, who had ridden hard all night from Washington to deliver to Tyler a letter, a document of history: President William Henry Harrison had died two days earlier, just thirty-one days after his inauguration. Tyler would be the first of nine Vice Presidents who have succeeded to the office. Unaware that "Tippecanoe" was seriously ill, Tyler had been taking his ease at his home; piqued that he had been given no share in doling out patronage, he had simply left Washington and returned to Virginia. The first "accidental President" raced back to Washington by horseback and chartered steamboat to take the oath of office in a hotel room.

His succession to the presidency was by no means automatic. When strong opposition in Congress attempted to declare him "acting President," Tyler successfully argued that the Constitution made him *the* President, empowered to assume the office for the full term. With that, he promptly established the first wave of his large family, along with several domestic slaves, in the President's House. With his resolute stand, Tyler set the immutable precedent for the instant and orderly transfer of power at the death of a President, the guideline followed in four assassinations, three natural deaths, and one resignation.

While some Vice Presidents might have been less staunch than Tyler, Theodore Roosevelt was definitely not of that ilk. In 1901 he was vacationing with his family in the remote Adirondacks, reassured that President McKinley was recovering from shots fired point-blank by an anarchist at the Pan-American Exposition in Buffalo. Then a grim telegram arrived. The President had suffered a relapse; the Vice President must "lose no time in coming." A "carry-all" with three relays of horses and drivers rushed him thirty-five miles—six hours on a treacherous mountain road—to a special train that sped him to Buffalo.

At the train station later that day his wife, Edith, returning with the children to their Long Island home, was handed a terse, formal message: "President McKinley died at 2:15 this morning. Theodore Roosevelt." Her husband, Vice President for only six months, had been swiftly sworn in as the twenty-sixth President; not yet forty-three years old, he was the youngest President in history. (Fifty-nine years later, John F. Kennedy at forty-three would be the youngest ever elected.)

Not in the deepest recesses of her soul did Edith Carow Roosevelt want this move. She had counseled Theodore, governor of New York, to stay in the statehouse and then run for the White House. Why think about the useless position of Vice President? But when the 1900 Republican convention went wild for Teddy, he had no choice, and McKinley reluctantly concurred. At that, GOP boss Mark Hanna sourly said to the President, "Your duty to the country is to live four years from next March!" It was unthinkable that the popular McKinley might be in danger in that placid time.

Edith's life as First Lady began, unforgettably, "behind a hearse in a carriage drawn by four splendid black horses," with her husband, the new President, in plain view. In a letter to her sister, she acknowledged her fear: "I suppose in a short time I will adjust myself to this, but the horror of it hangs over me. I am never without fear for Theodore." Happily, her fears gave way to almost eight exhilarating years in a family White House free of horror, full of fun.

THE THUMP OF drums and cheers for home-state floats flood in from the street, penetrating the north windows of the White House; the sounds of victory can almost be heard over the din of the vacuum cleaners inside. Inauguration Day is a day of frantic action at the President's House. While the ceremonies are in progress, the domestic staff must see that all fifty-four rooms and twelve and a half baths of the family quarters are spotless and clean every inch of those that have been occupied. Last-minute odds and ends must be packed, newly emptied closets and cupboards scrubbed and sprayed, furniture moved as directed, beds made up with the prettiest linens, all pillows plumped, fresh flowers and niceties put in place, the incomers' clothes hung in the rooms designated for which family members, which guests.

Downstairs another team is readying food to follow the parade and precede the ball. The next day the new First Lady can say how she likes things done, but today it is done the White House way. And it is exhausting. Those early White House tradition-setters slipped up in this matter: the outgoing family should be required to leave the White House a day or so before the changing of the guard—put up in Blair House or the best hotel suite to allow this enormous task to be handled sensibly. Some First Ladies have been packing their bags as the parade formed, while the new First Lady's inaugural-ball gown was waiting to be hung.

As it is, the staff race against the clock on Inauguration Day. After four, or eight, years of serving one family, mastering their likes and dislikes, they will be expected to adapt, between breakfast and lunch, to who knows what changes—maybe for the better, maybe not. Head Butler Fields, with the experience of serving four families in his twenty-one years, offered a basic rule for those who serve: "One is lost if one isn't quick to adjust."

The relationship between the mansion staff and the First Family is unique: the servants come with the house. And since they are not easily dismissed, it behooves both sides to get along. Jacqueline Kennedy described that symbiosis at its best in praising Chief Usher West: "He gave presidential couples the one thing they missed most in the maelstrom of the White House—a sense of tranquility."

The twenty-second amendment to the Constitution limits a President to two terms and his West Wing staffs leave with him, but there are no term limits for the scores of employees who work in the mansion, closest to the family. Some stay for decades. They were there before the new family arrives, and they will be there after that family goes. A White House gardener may see the trees flower long after the First Lady who ordered them has faded from memory; a houseman will have moved the same furniture around so often he knows where it won't fit.

On the other side of the equation, a new First Lady meeting the Executive Mansion staff wonders who among them will not give up their attachment to the departing family. Chief Usher West stated bluntly, "As time-clock government employees, the domestic staff were bureaucrats pretending to be old family retainers. And you simply can't be both at the same time." He observed that the household workers established "their own code, a subtle 'We'll be here after you're gone' attitude."

The new First Lady "will search their eyes, looking for loyalty." She knows that behind the noncommittal faces these people who keep the White House ticking make judgments and have favorites; some of them assume a proprietary attitude, disapproving of the new family and their new ways in the old house. In his twenty-eight years serving five presidential families, West saw each First Lady quietly "examining, scrutinizing, testing every employee. 'Are you with me?' is an unspoken question—and somehow the President and his wife can always tell if the answer is 'No,' or even 'I'm not sure.' The White House employee whose loyalties remain with the previous administration rarely survives long." (Perhaps something of that nature was at work when Laura Bush, going into a second four years, dismissed the chef hired by Hillary Clinton.) Those who can make the transition through several changes of families are a special breed; working within the most politically charged dwelling in the country, they have learned to master psychology and minimize politics.

But they make judgments, and sometimes they write books. In hers, Lillian Parks called the Hoover term "my most difficult" in thirty years of White House service. Utter silence and invisibility were required of servants; at the sound of the bell alerting them that the President or First Lady was approaching, they had to scurry pell-mell into the nearest closet. Even when the Hoovers dined alone in the cavernous State Dining Room, the butlers stood stiff as sentries, awaiting Mrs. Hoover's hand signals. Through four years, Lillian wrote, "the President never spoke to me, not even at the Christmas party." Then change came overnight. "Suddenly I was confronted by a President [FDR] who spoke jovially to the servants and wanted to know exactly what I did." And Head Butler Fields, balancing a loaded tray, was astonished when Eleanor Roosevelt insisted that he share the elevator with her.

Bess Truman, who seemed unbending, even dour, in public, was the First Lady the White House staff liked best of all. West found her wit "dry, laconic, incisive and very funny . . . the President and Margaret would howl with laughter at the dinner table." Who would have thought that Bess Truman was a rabid baseball fan? But she had been a tomboy who used to play second base herself. Even more, Bess showed concern for the staff as individuals—she once turned up to cook Thanksgiving dinner for a doorman whose wife was ill. Though her Gates-Wallace family were leading citizens of Independence, with a postcard house, servants, and deference, Bess, said the chief usher who worked with her every day, never seemed to feel "above" anyone. By contrast, he characterized Eleanor Roosevelt, the humanitarian, as "formal and distant with her staff" and at times quite brusque if asked to repeat an order, a problem compounded by her illegible handwriting.

When J. B. West offers comments about First Ladies—that Bess was warm and Eleanor cool, that Mamie was commanding and Jackie a conspiratorial friend—they are reliable. And a White House butler's unique experience—the impassive mien conceals sharp eyes, and an alert mind registers every nuance of the scene around him—holds lasting interest. That was proven again in 2004, when a dramatization of Alonzo Fields's memoirs—his quiet reflections and private judgments—had a successful run in Washington's historic Ford's Theater where the premier box will always be occupied by the ghost of Abraham Lincoln.

The departure of a First Family who has been enjoyed, even loved, by the men and women who see the White House's members as themselves, away from the public eye, can be a sad moment. For years those cooks and butlers, maids and gardeners, ushers and electricians, have served that family, sharing their private pleasures and sorrows. Then that First Family walks out of their lives, and once again there is anxiety about the new arrivals. "It can be as sudden as death," mused Fields. "And then we start all over again."

three

Growing Up in Headlines

———⟫•⟪———

No CHANGEOVER COULD be more discombobulating than that following the shocking assassination of President McKinley, just six months into his second term, and the arrival of the Theodore Roosevelts—the quietest of presidential couples followed by the liveliest of presidential families. Six boisterous children, not to mention the President himself, took a bit of getting used to.

But it wasn't long before the hardier staff members joined in the jollity. There was, for example, the memorable caper when young Archie was in bed with measles. Brother Quentin knew just the thing to speed his recovery: with Quentin pulling and his White House accomplice pushing, Algonquin, a small Icelandic pony, reluctantly squeezed into the elevator, uncertain about the whole thing. Icelandic ponies, though tiny, are not bred to ride in elevators, certainly not in the President's House, but Quentin was sure that a visit by Algonquin was just the thing the doctor didn't order to cheer Archie up—which it did, to gales of laughter.

The White House Gang, as the four Roosevelt boys styled themselves, discovered great possibilities in their new home. They "borrowed" the biggest metal trays in the pantry for sledding down the steep stairs to the State Dining Room, once nearly upending a waiting ambassador; they skated in the East Room and found the high ceilings and long unbroken sweep of the ground-floor hall ideal for stilts and bicycles. But when they pulled out water pistols in the East Room, the chief usher was forced to declare it off-limits. That didn't stop Quentin and coconspirator Charlie Taft from lobbing spitballs at Andrew Jackson's portrait, which drew a rare reprimand from the President. (Charlie, whose father was secretary of war and would be the next President, invented his own White House mischief: slipping underneath the dinner table, he would tie the guests' shoelaces together.)

On the White House grounds the gang pitched a tent, waded in the fountain, staged war games, and raced anything with wheels. After dark the older ones followed the lamplighter around Lafayette Square until he was out of sight, then shinnied up the poles and snuffed the lights. And they roughhoused with the biggest kid of the lot—the President of the United States.

The Roosevelt White House made a fine ark for the young Noahs' menagerie—ponies, dogs, cats, raccoons, white mice, guinea pigs, frogs, lizards, parrots, a bear, and a macaw named Eli Yale. White House staff watched where they stepped and averted their eyes when the President fed sugar to the pet kangaroo rat at the breakfast table.

Alice, the eldest, was devoted to the special pet she had bought for herself: a small green snake named Emily Spinach. (Her aunt Emily was very thin; spinach is green.) Alice took it everywhere, unconcerned—or perhaps pleased?—that her hostesses at parties and weekend visits were not especially charmed by Miss Spinach. "She got lost in the folds of the drawing room curtains at one house," Alice recalled, "and shed her skin under my bed at another." When Emily was found dead outside Alice's window, she suspected "foul play." Her friends cheered.

Why on earth would Alice flaunt a snake? Sheer exhibitionism, seeking attention—hopefully from her father. For Alice, having the nation at her feet was not enough. It was the love of the father she idolized that she yearned for, and her outrageous behavior was one way of getting him to notice her. A touching entry in her diary offers the clue: "Father doesn't care for me, that is to say one eighth as much as he does for the other children. We are not in the least congenial." TR was an outdoorsman who liked to hunt big game; Alice was a party animal. Understandably, their relationship was uneasy: he had turned her over to an aunt for the first three years of her life, and somewhere deep in his psyche there might have been resentment toward this child who had cost the life of his adored first wife after childbirth when she was only twenty-three.

As for the boys, there is no need to scrutinize their hijinks for deep psychological roots: they simply discovered in the White House endless opportunities for uninhibited fun. Their father reveled in their inventions, rarely imposing even the slightest discipline.

Discipline was what Miss Arnold, young Quentin's teacher, had in mind when she suggested in a letter to his parents that a measure of it might be in order for the unruly eight-year-old. By return mail came a tart letter from the President, countering that "it would be well" to administer the discipline at school and assured her that he had "no scruples against corporal punishment." (Not that he imposed any.) He ticked off Quentin's offenses, "such as dancing when coming into the classroom, singing higher than the other boys, drawing pictures rather than doing his sums . . . which look as if they could be met by discipline in school and not by extreme measures taken at home." Having had her knuckles cracked by the chief executive, Miss Arnold may have sought another line of work.

Quentin would convert his mischief into boldness in the First World War; a fearless fighter pilot, he was downed in combat with a German plane. His mother was philosophical in her grief: "You cannot bring up boys as eagles and expect them to turn out sparrows."

In their almost eight years there, the rollicking Roosevelts had a greater impact on the White House than the White House had on them; they transformed it from Executive Mansion into the Roosevelt family home and gave the nation more laughs than the comic strips' Katzenjammer Kids.

FOR ALL ITS perks and prestige, the White House cannot supply a President with the respite from problems and happiness that children bring. We think of Lincoln as a somber President, despite his sharp wit and homespun anecdotes, but in his first year there the White House was filled with fun and laughter, games and escapades, devised by his young sons. He wrestled with them; he paid the five-cent admission for the show they staged in the "theater" he had built for them; he allowed them to sell lemonade to job-seekers jamming the halls (it's a safe bet that few turned them down) and was amused when eight-year-old Tad (Thomas) bombarded the door of the cabinet room with his toy cannon. Lincoln dismissed the grumbling about his sons' behavior. "They'll get pokey soon enough," he argued, probably thinking of his eldest son, Robert, a Harvard student who had indeed become "pokey."

Then the unthinkable happened: their eleven-year-old Willie (William Wallace) contracted what was probably typhoid fever. While his parents kept watch, helpless, at his bedside, Willie, their favorite child, the son most like his father, slipped away. It was the Lincolns' second devastation—they had lost little Eddie, their second son, when he was not yet four. Tad was inconsolable; Mary was unable to function. After a time, a desperate Lincoln, pointing out a nearby insane asylum, told her, "Mother, you must pull yourself together, else you will land over there." The President, already burdened with the fate of the Union, was forced to bear the tragedy within himself; Willie's death surely marked the beginning of Mary Todd Lincoln's long descent into madness.

After the loss of Willie, the Lincolns indulged obstreperous Tad even more, undoubtedly trying to help him—and themselves—

through their unbearable grief. The head of the domestic staff, Colonel William Crook, reminisced years later, "We all liked to see the President romp up and down the corridors with Tad, playing horse or blind man's buff. It seemed good to have him happy, and he was happy when he was playing with the boy." Mary was equally permissive, even when Tad and a friend hitched his pet goat, Nanny, to a dining room chair and trotted into a state parlor where she was greeting distinguished visitors. To help Tad play soldier, the secretary of war commissioned him a "Union lieutenant" with his own miniature uniform, and Tad, flaunting his new power, dreamed up even bolder stunts.

Though Tad brought a moment's relief from the grim events surrounding the White House, some members of the staff mumbled that the boy needed tighter reins. The Lincolns, however, did not believe in discipline, probably compensating for the President's own bleak childhood. A possible reason for spoiling Tad may lie in historian Matthew Pinsker's description of the boy as "a slow-developing child, hindered by a speech impediment. . . . [At age seven] he did not dress himself, nor could he read or write much until well into his teenage years." Tad's tutor, a government employee, got nowhere with him, while Willie, such a different child, was already showing his father's gift for writing and love of books.

But Tad was touchingly tenderhearted. When Jack, the turkey being fattened for Christmas, became his pet, Tad's distress was so great that his father granted the bird a pardon—a custom that continues today. Another time, Tad, moved by the plight of a poor woman waiting to entreat the President to free her husband from prison, ran to his father and clung to him until he agreed, leaving the petitioner, Tad, and Colonel Crook in happy tears.

It comes as a surprise to find powerful fathers so permissive in the nineteenth century, an era when children did not rule the roost, yet Ulysses Grant, the general who had commanded troops in battle, was a marshmallow when it came to his youngsters. Like the

Lincolns, the Grants did not believe in discipline; they delighted in indulging their four. The two older Grant sons were away at school—Fred at West Point, Buck at Phillips Exeter Academy. But eleven-year-old Jesse, almost as mischievous as Tad Lincoln, took over the White House from the roof, where he turned his telescope on the night skies, to the tool shed, which he converted into a clubhouse for the secret society he formed with eleven pals. (Fifty years later, aging members of Jesse's secret club gathered for a unique reunion, reminiscing about the days when they had the run of the President's House.)

Pert thirteen-year-old Nellie, the apple of her daddy's eye, became the nation's darling from the first day: as her father delivered his inaugural address, she came up to him and took his hand. Not until someone placed a chair close by him did she let go. In their second White House year, Nellie was enrolled in the fashionable Miss Porter's School in Connecticut, but after only one day she sent her parents a telegram begging to come home, promising to help her mother with White House receptions.

Her indulgent parents agreed and didn't seem to mind that Nellie spent more time flitting about Washington in her smart phaeton and starring at the best parties than helping her mother at White House events. She was even allowed to attend all-night dances, called "Germans," which started at eleven and lasted until five in the morning. Such license brought the stern disapproval of very proper journalist Mary Clemmer Ames, who *tsk-tsk*ed that young Nellie's health might suffer from "the wild, unhealthy excitement through which she whirls night after night." Wild excitement was just what Nellie thrived on.

Then Jesse, sent to a boarding school in Pennsylvania, borrowed Nellie's tactics, immediately pleading, "I want to come home!" His mother felt he must stick it out longer, but the soft-touch President wired back: "We want you, too. Come at once." In his undistinguished later life, Jesse made an attempt to follow his father into the

White House, but the famous name was not enough. Julia Grant may have realized that Jesse needed a measure of discipline.

SIX DAYS AFTER moving into the White House, Hillary Clinton, anxious about how to give her adolescent daughter, Chelsea, a normal life in the glare of the White House, paid a private call on the woman who would best understand the problem. A sympathetic Jacqueline Kennedy Onassis did not minimize the challenge. When she was in Hillary's shoes, she said, she had felt that same anxiety— "I wondered, how are we going to raise a family in this place?"

In that quiet meeting, Hillary wrote in her memoirs, Jackie warned her that it "would be one of the biggest challenges Bill and I faced. We had to allow Chelsea to grow up and even make mistakes while shielding her from the constant scrutiny she would endure as the daughter of a President." Jackie's advice was specific: "You've got to protect Chelsea at all costs. Surround her with friends and family, but don't spoil her. Don't let her think she's someone special or entitled. Keep the press away from her if you can—and don't let anyone use her."

Back in 1869, First Lady Julia Grant had faced that same concern with her Nellie. An indifferent student at best, Nellie was kept after school one day until she completed her homework; three times the White House carriage called for her before she was allowed to leave. The next morning Julia called on the headmistress at the prestigious academy, not to protest but to thank her for not bending the rules for the President's daughter: "Teach her she is plain, simple Nellie Grant, entitled to no special privileges." Julia, concerned that the fawning staff and emoluments of the White House could lead to a princess complex, tried to prevent it, with mixed success.

Chelsea proved immune to the princess syndrome. Staying out of the camera's eye, she caused not a ripple of adverse publicity through four years of high school and four more at Stanford University.

Later, in her two years at Oxford, the British tabloids unearthed nothing racier than Chelsea snuggling with her boyfriend. Then she waltzed into an international consulting firm in New York at a salary estimated to top $150,000 a year and, like a hot young pitcher, may have received a signing bonus of $10,000. She arrived with a cum laude record from the college league—and a pair of well-known coaches.

"Whatever else you do," Jackie Kennedy once said, "unless you raise your children well, I think you have failed." She met her own standard. Though she had every kind of help for the children—nanny, servants, and the watchful presence of the Secret Service—Jackie was a hands-on mother, and unlike the Lincolns, she did not hesitate to use discipline. In fact, the White House was less a problem for young Caroline and John than for teenage children; Caroline came with a flock of cousins and playmates, while Chelsea was an only child, a stranger to Washington—and a teenager. But she navigated the shoals without incident.

For young children the White House is full of excitement with its endless crannies to explore and a private parade when presidents and kings come calling—bands playing, soldiers marching, flags flying in your backyard. At one state welcome little John Kennedy, watching from the balcony, launched a toy airplane that scored a direct hit on a soldier below, to the delight of camera crews. (That soldier has a unique tale to tell, if not a Purple Heart, whenever grizzled veterans swap combat yarns.)

Over the years the offspring most ambivalent about living in the White House have been the Presidents' teenage daughters, forced to deal with its heady mix of glamour and restrictions, of too much publicity and too little privacy, of hurtful criticism and fawning insincerity. But most of them look back on it as a great experience.

Reflecting on the downside of life in the White House, Susan Ford, now forty-eight, twice married and the mother of two daughters, muses, "If I had to do it again at that age—seventeen—I wouldn't want to. But at this age I would travel more, be more involved in watching meetings, in how government works." Most of the Presidents' daughters would share the view of Margaret Truman: "Living in the White House is a unique experience—a fantastic compound of excitement and tension and terror and pride and humility. Above all it is a historic experience."

Almost forty years after leaving the White House, Lynda Johnson Robb and Luci Johnson Turpin, now striking beauties at sixty-three and sixty, look back in gratitude at the five memorable years that defined their lives. Shortly after moving in, sixteen-year-old Luci declared dramatically, "I have been robbed of my youth, my private life," but she quickly accepted the love/hate relationship between First Daughters and the White House: "You can adjust—or you can adjust!" Now Luci, mother of four, grandmother of nine, married for the second time, and head of the LBJ Holding Company, talks about those years: "Initially, I saw how the White House limited my life. I saw it as a source of irritation. I felt I was living in an imposing museum, a public fishbowl and a prison. Then I came to see it as the golden opportunity to be an eyewitness to the major events of my youth, to meet the movers and shakers of my time. I began to realize this extraordinary gift of opportunity and told myself, 'Don't blow it!' "

Lynda, three years older than Luci, was a college student when the family was thrown into the White House amid national mourning. From the outset she was realistic about her new role. "I have no illusions," she acknowledged not long after moving in. "I know that people are not applauding me—they're applauding the President's daughter." She soon conceded that living in the White House "does have its good points—having the spotlight on you really matures you."

Now both daughters remember those years with gratitude. Lynda tells about the day former President Truman unexpectedly joined the family for lunch. While Lady Bird was dismayed that they were having hash, Lynda seized the opportunity to ask him directly how he had felt about dropping the first atomic bomb—"He said he was sure it was the right thing to do, and he got a good night's sleep afterwards." That family lunch with the former President, she recalls, was "one of those times I knew how lucky I was to be there." Always history-conscious, Lynda made notes of important events she witnessed: "Some of it was trash," she says now, "but there were some nuggets."

With the malleability that marks the White House, the top-floor solarium that had been Caroline Kennedy's schoolroom was quickly transformed into the Johnson girls' teen room, replete with Coke bar, record player, TV, friends, dates—and no Secret Service. More than most White House children, Lynda and Luci lived all over the mansion. Luci labored over algebra in the Treaty Room, and Lynda pursued history in the Yellow Drawing Room, the oval centerpiece of the family quarters, the room FDR used as his study and Winston Churchill commandeered during his wartime visits. An anxious Lady Bird preferred the daughters to use less historic desks.

With a touch of nostalgia, Chief Usher West recaptured those days: "They kept the staid old halls jumping. Noise and laughter, dates and dramatics echoed through the historic rooms. With young romances on the third floor, Congress up and down the stairs, and meetings all over the place, the Johnsonization of the White House didn't involve the decor, it involved people. The house looked the same, but slowly the Kennedy style was erased, just as the Eisenhower style had been erased previously."

Luci (like Jefferson's daughter Mary, who became "Maria" in Paris, Lucy morphed into "Luci") shared a concern with every White House daughter: is it herself or the glamour of the President's House that attracts the new friend? "If I am liked," she said, "I want it to be for me as a person." With its aura of power, the White House can twist friendships—sometimes friends pull back, reluctant to seem to be cozying up; others try anything to squeeze into the inner circle. It's not always easy to tell the difference.

It may be hardest for the college student, thrown into classes of equals where the reaction may be hostile or pandering. In her first September as the President's daughter, Margaret Truman, entering her junior year at George Washington University, a few blocks from the White House, had hoped to continue as just another history major—wishful thinking for the first unmarried First Daughter since the Woodrow Wilson girls twenty-five years earlier. She immediately felt the full force of her new identity: "Boss' Daughter Great Catch for Anyone" shouted the front page of the student newspaper. The "great catch" was horrified, and her father, predictably, was enraged.

Contrary to the student editors' notions, Margaret found that the White House restricted her social life, as later daughters would agree. "I ask you to consider the effect of saying good night to a boy at the door of the White House in a blaze of floodlights with a Secret Service man in attendance," Margaret fretted in her memoirs. "There is not much you can do except shake hands and that's no way to get engaged." And if she stayed out later than her watchful father approved, he ordered the Secret Service to bring her home.

"The Secret Service makes me nervous" ran the line of an old Broadway tune, and every marriageable White House daughter would add her amen!

Susan Ford, entering the White House as an unspoiled seventeen-year-old with no exposure to public life, now recalls, "For a teenager, the White House seemed like a cross between a reform school and a convent." The Fords had been in their new roles only six weeks when her mother underwent an operation for breast cancer in October 1974. A few days later Susan, despite her anxiety and fears for her mother, stood beside her father, acting as hostess for a diplomatic reception.

Tall, blond, and glamorous in a chiffon gown, she looked every inch the part, but, "I was not prepared to be the unofficial First Lady. I had never worn long white gloves before. I was scared to death that I might do something wrong—like call someone by an incorrect title or step on his toes while dancing." She remains grateful to her mother for instilling the "manners" that prepared her for meeting heads of state and celebrities and coping with White House social demands. "I was nervous, but at least I had the basics. I could go from blue jeans to long white gloves and be back in blue jeans an hour later."

Growing up in the capital—she had lived her entire life in the same suburban Virginia home—made White House life easier for her than for some White House teenagers. "Luci and Lynda and I were raised in Washington," Susan points out. "We had friendships, we were part of the Washington scene." Still, she knew the White House only through her summers there—selling guidebooks. Without a campaign or an inauguration to help her get used to the idea, the new life at first seemed unreal. "When we first moved in," she said with a laugh, "I picked up matchbooks, note pads—anything that had 'The White House' or 'The President's House' on them. It was as if I was leaving the next day." For the Bush twins who had never lived in Washington, the solution in their father's term was to stay away. But Susan feels strongly that if a President's children prefer to stay out of sight, "it's their choice, and they should have the right to make that choice."

THE BRIGHTLY OUTGOING, nonidentical twins, Jenna and Barbara, would welcome Susan's opinion. They want no part of the White House or its majesty; they have preferred to be with their parents at their Texas ranch, or occasionally Camp David, when away from their respective colleges. In 2004 Jenna graduated from the University of Texas in Austin, the city she had known from her six years in the governor's mansion, and Barbara added to the family record as the fourth-generation Bush to graduate from Yale. The President once said his biggest fear about his daughters was that they would not have a good time in their college years, but clearly they did. Chatting with visitors to his campaign website, they declared, "We were able to live normal lives hanging out with great friends . . . college was so much fun," and traveling in Europe with friends after graduating was "a blast."

However, staying clear of the Executive Mansion does not guarantee staying clear of publicity. In 2002 Jenna was arrested twice in five weeks for underage drinking, once using a fake ID at a favorite hangout in Austin. She was sentenced to community service. Barbara made the New York gossip columns doing the trendy downtown club scene, and when she took to the dance floor to join the performer, a Japanese dancer, with a lot of body action, her pals threw flowers and cash.

A burst of criticism met Jenna's flouting the law and Barbara's flaunting her moves. Laura defended them, insisting that "they just want to do what every teenager does," and when the twins were reported to be deliberately giving their Secret Service agents the slip, Laura—differing with their father—insisted that the agents pull back to give them more space. Like the Lincolns and the Grants, Laura Bush parents with indulgence more than discipline. But former First Lady Barbara Bush, when asked by Jay Leno on *The Tonight Show* if the press had been too hard on her granddaughters, replied with her legendary candor: "That was not what they should have done. It was stupid. They are smart girls—they knew they shouldn't have done that." The First Grandmother's forthright opinion reminded viewers why she led popularity polls for years.

As their father's reelection campaign got under way, the twins, who had previously expressed no interest in politics, decided to put aside their antipathy to the spotlight and help in the effort, targeting young audiences. Their public debut was a bland interview in *Vogue* featuring the two in glamorous ball gowns by top designers, their first foray into classic sophistication. The gowns were not added to their own wardrobes—theirs is the edgy in-the-moment, lace with jeans, bare-midriff school of fashion.

Not to be outdone by John Kerry's daughters, the twins, nearing their twenty-third birthday, were featured speakers at the Republican National Convention. Jenna, doing most of the speaking for the two at the podium, delivered lines (obviously written for her) that met with uneasy reaction among the overwhelmingly conservative delegates. She mischievously confided to the crowd that her grandmother is "just not very hip—she thinks *Sex and the City* is something married people do but never talk about." Explaining their absence from the White House scene, Jenna went on, "We spent the last four years trying to stay out of the spotlight. Sometimes we did a little better than others. We kept trying to explain to Dad that when we were young and irresponsible—well, we *were* young and irresponsible."

Yet on her first day of campaigning, Jenna peevishly stuck her tongue out at the media as she waited for her father in his White House limousine. How could she think the childish act would not be all over the newspapers and television—or was it calculated? Laura Bush's mild reprimand was, as she told it on CBS, "Maybe you should work on your issues of impulsiveness or something." Now twenty-six, the Bush twins are emerging as independent adults. Jenna, after teaching third grade in a Washington charter school, is working as an intern with UNICEF in Panama; Barbara is on staff at the Cooper-Hewitt Museum in New York. Both are regular grist for the gossip columns.

STAYING OUT OF sight was taken to a new level by the Reagan off-spring. All four were present at their father's inauguration in 1981, which was the last family photograph in Ronald Reagan's two terms in the White House. Eight Christmases and many memorable events came and went, but the only Reagan child who participated was Maureen, the President's daughter by his marriage to actress

Jane Wyman. This very political daughter loved the White House, its power and excitement, and stayed there for long periods as she tried, unsuccessfully, to carve out a political career of her own. (When she sought Republican support for her dream of running for the Senate in California, her father stayed neutral, but it was no secret that he took a dim view of her coattails candidacy.)

At the other extreme was Ron and Nancy's daughter, Patti. No presidential offspring has ever been as defiant in shunning the White House. Turning her back on her parents, she rejected the Reagan name; she chose instead to seek individuality—if not anonymity—under her mother's family name, Davis. The rebellious daughter vented her anger against her parents' indifference by turning semihippie and publicly speaking out against Reagan's policies. Michael, the older son whom Ron and Jane had adopted, brooded that he was the rejected child because he was adopted. Had he looked closely at the dysfunctional family, he would have realized that all of the Reagan children were ignored—the President and his Nancy were a couple complete in each other; they treasured their closed togetherness in the White House.

The youngest child, Ronald Prescott Reagan, seemed to accept his parents' distance with equanimity. In a television documentary about his father, Ron observed, "In our house, there were Nancy and Ronnie—and then there were the rest of us." Staying clear of his father's world, he sampled Yale, danced with the Joffrey Ballet, and remains an activist for the arts; he now works in nonpolitical television. It may be no coincidence that he and his wife, Doria, a clinical psychologist, chose to live in Seattle, a world away from both the White House and Beverly Hills. And he is the only Reagan offspring who has not written a book.

Following his father's death in June 2004, Ron dropped his self-imposed silence on political issues. At President Reagan's burial at his presidential library, in the serene spot he and Nancy had selected, the son paid a strong farewell tribute. He pointedly declared that his father, "an unabashedly religious man, never made the mistake of so many politicians—wearing his faith on his sleeve to gain political advantage." (Ron terms himself an atheist with "sympathies" for Buddhism.) It was a direct thrust at President Bush. The following month came the dramatic moment at the Democratic convention: the bearer of the ultimate Republican hero's name was speaking to the nation, in prime time, against President Bush's position on stem-cell research.

Patti at last made peace with her parents and herself. As her father slipped into the mists of Alzheimer's, she whispered to him, "I'm here now. I'm not leaving"—the words she longed to have heard from him. Though too late to have meaning for him, the moment was healing to the estranged daughter, who now lectures on forgiveness and family; shortly after his death her book, *The Long Goodbye,* was published, a sentimental story told by the prodigal daughter, a pole apart from her earlier antiparent screed. Son Michael had long since closed the chasm between himself and his father; his syndicated radio talk show not only embraces his father's conservative policies but toughens them.

HAS ANY CHILD found the White House less congenial than Amy Carter did? The photograph in *Time* magazine brings it all back: nine-year-old Amy on her first day at a time-worn public school, walking past the press horde held in check by a rope. She is wearing a new winter coat, pants, and shiny new shoes; her strawberry blond hair spills out from a stocking cap, and she carries a new Snoopy schoolbag. She could be any other fifth-grader, except for her downcast head and a little face that is both sad and unyielding. Her mother, the new First Lady, walks a few paces behind, making Amy the sole focus of the cameras. *The Washington Post,* going a bit overboard, described her as "forlorn—a baffled and beleaguered public figure."

Enrolling Amy in this 108-year-old inner-city school—60 percent of its 217 students were black, 30 percent were foreign-born, mostly children of embassy workers—demonstrated the egalitarian values of the new President, a chief executive who carried his own garment bag and would be known officially as Jimmy.

The magazine's cover pictured a different Amy, sweetly at ease with her dog, Grits, in the sheltering arms of the White House. Out of the public eye, she was a lively kid who roller-skated down the long White House halls, played in the treehouse built for her on the South Lawn, and loved the private screening room where she could order up her favorite movies and snacks. (That week it was Disney's *Freaky Friday* for herself and a schoolmate, the daughter of a Chilean embassy cook.) She went places, met personages, and was hugged by the Pope when he called at the White House in 1979.

Exciting, and yet . . . Amy, the youngest White House child since Caroline and John Kennedy and, before them, Teddy Roosevelt's unruly gang, was loved but lonely. Rosalynn, eager for Amy to feel the history of their new home, walked with her through the rooms with a guidebook "matching the names with the pictures on the walls." The Carters made an effort to be normal parents, but normality isn't easy when you must travel everywhere with security guards and a press pool and there is always a cluster of—hopefully— friendly bystanders. Still, they tried. When Amy was one of a handful of boys and girls in a summer program for gifted children, Jimmy and Rosalynn attended a program celebrating the group's ethnic diversity. (Amy read a myth about the sun and the moon.) Another time they drove out to the Maryland suburbs to hear Amy in a violin recital. Their eagerness to include Amy in their White House life backfired when she attended a state dinner for the President of Mexico, and was seated next to the foreign minister, who was chagrined to be seated by a child—especially one who read a book throughout the meal. Replying to the predictable criticism, President Carter countered, "We always read at the table when we were growing up." But as Dorothy memorably observed, "Toto, we're not in Kansas any more." It was not an experience to build confidence in a little girl.

But White House parents are always stretched thin, and Amy's three brothers were much older and married, two with babies. Her doting grandmother, the irrepressible Miss Lillian, was far away in Plains, Georgia. Amy had lost the easy independence of her small-town childhood. Without doubt, the White House left a permanent impression—perhaps a scar—on young Amy's psyche.

Away from her family, at Brown University she became a political activist well to the left of her father, protesting CIA covert actions and South Africa's apartheid—and was arrested with much media attention. Putting more time into passive resistance than active studying, she in effect flunked out. Twice she canceled weddings (to different grooms), then settled down to develop her special talent; she holds a master of fine arts degree from Tulane University and has illustrated some of her father's books. Unlike other White House daughters, Amy has never once emerged in a public role or spoken to the press.

No White House daughter has ever been more press shy than Amy Carter; thus it followed that her wedding would be a ceremony as different as possible from White House traditions. She married James Wentzel, a website designer, by the pond at her grandmother's secluded house at Plains. Vows were exchanged under an arbor she devised from her old swing set, intertwined with vines; she baked the wedding cake, declined to be "given away" and allowed no photographs to be published. Marching to her own drummer, Amy Carter was still reacting against the demands and strictures, the pageantry and publicity that come with the honor of the White House. She has one son; the marriage ended in divorce.

Amy was the exception among First Daughters. Most of them found more pleasure than pain in the White House—especially in retrospect. Fifty-five years after the Woodrow Wilsons moved into the Executive Mansion, Eleanor, who at twenty-three was the youngest of three daughters, told about her first morning: "It was fun to wake up in the White House, to jump out of bed, and laugh with Jessie [her sister] because we didn't know where the family dining room was." But this daughter also felt fear; she remembered crawling under her bed, facedown on the carpet, sobbing, "It will kill them. It will kill them both." A daughter's fears were prophetic—her mother died the following year (from a kidney disease), and her father would suffer a damaging stroke in his second term.

The eldest Wilson daughter, twenty-six-year-old Margaret, was unique among her White House peers. Dreaming of a singing career, in World War I she gave concerts for the Red Cross, reaping more than $10,000 in her first performance—the cause was good and she was the President's daughter. She cajoled her father into letting her go to France to entertain the troops and won high praise for cheering homesick soldiers. But after her father left the White House, physically and politically powerless, Margaret's aspirations for a career in music evaporated.

Her life then took an incredible turn—she fell under the spell of a widely known Indian swami, Sri Aurobindo, joined his ashram in Pondicherry, India, and adopted the name "Nishta." There she wrote a book with philosopher Joseph Campbell (later made famous by Bill Moyers on PBS) whose mantra "follow your bliss" became Margaret/Nishta's own. She never returned home. In 1944 she was buried in the Protestant cemetery in Pondicherry, her simple headstone chiseled with both "Nishta" and "Margaret Woodrow Wilson," the summary of her two lives.

DESPITE ITS STRICTURES, the White House can be fun. Who else could invite pals for a sleepover in the Lincoln bed (the mattress was lumpy), as Margaret Truman did, listening for Abe's ghost, said to wander his old familiar corridors by night. And in the White House cinema Margaret called for *The Scarlet Pimpernel* seventeen times—she was smitten by its star, elegant British actor Leslie Howard. Could it be coincidence that her future husband bore a striking resemblance to the leading man she had found irresistible in the White House screening room?

At minimum the White House sends its daughters away with a lifetime of memories; often it launches them on a rewarding public trajectory, writing, speaking, serving on civic boards. However they felt about the White House when they lived there, it propelled them into lives of achievement and service, and they speak gratefully for the push.

Three daughters began to write for long-established magazines while still in the White House—Lynda Johnson for *Ladies' Home Journal,* Julie Nixon Eisenhower for a revival of *The Saturday Evening Post,* and Susan Ford for *Seventeen.* Though they were generally qualified, the White House was a very helpful launchpad. Margaret Truman, Julie, and Susan went on to write well-received books. (Margaret holds the record—in addition to editing her father's memoirs and writing the life story of her reticent mother, she is the prolific author of murder mysteries set in such intriguing places as the White House and the CIA.)

Little Caroline Kennedy grew up to earn a law degree, have three children with her husband, Edwin Schlossberg, and coauthor books on the First Amendment and the issue of privacy, surely influenced by her mother's concerns in and out of the White House. She edited an anthology of Jackie's favorite poems following her death, borrowed her father's Pulitzer Prize book title for a new *Profiles in Courage,* and produced *A Patriot's Handbook,* a collection of poetry and songs that sing the praises of America. Caroline quietly lends her enormous prestige to good causes and stays out of unedifying headlines—all in all, the ideal White House daughter.

Not only daughters have become authors. Two generations of Eisenhower sons, clearly influenced by their White House experience, became distinguished military historians—John Eisenhower, who followed his father to West Point, and his son, David, now at work on the third volume of his biography of his grandfather, have both received high praise for their work.

WHEN THE PRESIDENT is also a father, one role or the other will likely be shortchanged—and the children often come second. FDR's sons had to make an appointment to talk privately with their father, and their peripatetic mother was scarcely more available. Eleanor later recounted the time one of the boys urgently wanted a personal discussion with his father—and as he talked, FDR continued to read through a stack of papers. Afterward, the boy angrily told his mother, "Never again will I talk to Father about anything personal!" Eleanor, calling it a slap in the son's face, reflected, "I sometimes wonder if the American public understands the toll the White House takes on a family's life."

A President-father can find himself more symbol than human. Eleanor Wilson McAdoo recalled her hurt in realizing that Woodrow Wilson "was no longer my father. These people, strangers, who had chosen him to be their leader, had claimed him. He belonged to them. I had no part in it. I felt deserted and alone." Many decades later Steve Ford, now a film actor who is also in great demand as a motivational speaker, had much the same reaction: "You can tell that people are looking through you to some image they have of your father, the President. You feel almost nonexistent, like you are standing in front of a symbol."

Eleanor Roosevelt, in her sympathy for her sons, might have given a thought to the vexation their grown children caused their father. "The private lives of the Roosevelt children were on the spectacular side," observed Bess Furman of the Associated Press (and, later, the *New York Times*) who covered the FDRs in the White House. Their escapades ranged from the younger sons' speeding tickets to multiple marriages (ultimately nineteen) among five children and assorted scrapes that produced a steady flow of gossip.

But when war came in 1941, all four sons served admirably, a source of great pride to their father as he led the nation through harrowing years of war and died within days of victory in Europe. Shortly after his father's death, Franklin junior, commanding a destroyer escort in the South Pacific, was awarded the Legion of Merit for chasing and sinking a Japanese submarine. It was a heroic action that would have made his father proud—and the father's approval would have made the son proud.

The devil-may-care FDRs were a U-turn from the Hoover brothers, who were as unsociable as their parents. Herbert junior, said Head Butler Alonzo Fields, was "like his father in that he did not appear to see others." The younger son, Allan, at Harvard Business School, did not take to life in the President's House. "If I don't get out of here soon I'll have the willies," he protested openly. While he was there, however, he ordered disruptive changes, replacing a butler whose walk he did not like and transferring another to a houseman's job, Fields recalled, "because Mr. Allan did not like his appearance . . . and that was that."

A PRESIDENT'S SON bears the double burden of living up to both the reputation and the expectations of the powerful father. Theodore Roosevelt, Jr., felt that "it handicaps a boy to be the son of a man like my father, and especially have the same name." Trying to carve out a political career of his own, he protested, "I will always be known as the son of Theodore Roosevelt and never as a person who means only himself." In the earliest days John Adams, who had nurtured John Quincy's career, harshly challenged his son: "If you do not rise to the head of your country it will be owing to your own laziness and slovenliness." Benjamin Harrison, the only presidential grandson to reach the White House, forbade his staff to mention the family connection, insisting, "I am the grandson of nobody."

Moving his three sons and four daughters into the Executive

Mansion in 1841, John Tyler, the first "accidental President," warned of the hazards that lay ahead. "Bear in mind three things," he admonished them. "Show no favoritism, accept no gifts, receive no seekers after office." His dictate should have included a fourth point: stay out of the headlines. From Eliza Monroe to the Bush twins, from the Adams sons to John Van Buren and the FDR boys, the annals of the White House are sprinkled with stories that embarrassed President-fathers.

The pitfalls abound. Consider the close call with a spicy headline narrowly averted by handsome Steve Ford, the youngest of the three Ford brothers: he was entertaining a young lady in the Queen's Bedroom when he suddenly heard the unmistakable voice of Barbara Walters, who was being shown around the family quarters by his mother. Exactly how he defused the French farce situation remains a secret, but television's celebrated journalist was unaware of the story she was missing.

Calvin Coolidge was one President who took action when he saw trouble ahead for his son. When John's performance at Amherst was sliding precipitously—his mind being preoccupied with pretty Florence Trumbull, the governor's daughter—the President dispatched his Secret Service agent, Colonel Edmund Starling, to serve as a combination babysitter and drill sergeant. "John was embarrassed and so was I," Starling recalled. "However, he inherited his father's sense of humor. He decided to make the best of it, and I had no intention of sticking his nose into a book and holding it there." He slept in John's room and "found some way of killing time" while his ward courted Florence at Mount Holyoke. At age fifty Colonel Starling sampled the easier side of a college education, and, yes, John married the girl.

Happily, more sons have come through the White House years unscathed than troubled. Two families with sterling records came with successive late-nineteenth-century Presidents—the four sons of Rutherford Hayes, followed by the Garfields' four boys. All were successful, and each of the Garfields was distinguished in his field. After serving in his father's White House, Major John Eisenhower became a fine writer, his true calling. The White House made a lasting impact on James Earl "Chip" Carter III, who lived in the Executive Mansion (on the payroll of the Democratic National Committee) with his first wife and baby son; witness his work with the Carter Center's international programs, his leadership in organizing young people's exchange visits and his present career in international business. But it was Jack, the eldest Carter son—who felt that his father was never really "there" for him in the White House years—who at fifty-nine jumped into politics, running for the senate in Nevada in 2006. Within one family, Amy and Jack embody the extremes in their reaction to the political spotlight: one shunning, one seeking.

President Tyler might thus have added a fifth point to his dictate to his children: do not tipple. One thread running through the lives of White House sons cannot be overlooked: the disproportionately large number who were alcoholics or scalawags, causing their parents public embarrassment and private pain. Over the years, the White House has been a pernicious influence on the weak; its power and privilege are a petri dish for self-indulgence and skewed judgment.

Distinguished as it was, the two-President Adams family had black-sheep sons in both generations, beginning with Charles, John Adams's scapegrace second son, who died of alcoholism three weeks after his father had moved into the White House. For John Quincy Adams, the shame of his brother's death must have imposed an even heavier obligation to live up to his father's expectations, to enhance the luster of the great New England name.

But the anguish of alcoholism—perhaps a genetic flaw—struck again in the next generation. After leaving the White House in defeat, John Quincy Adams summoned his eldest son, George, to Washington (the parents had remained in the capital) for what would surely be a stern dressing-down for his scandal-ridden, dissolute life. Journeying south by steamer, George vanished overboard, quite certainly a suicide. "No one knows the agonies I suffered," lamented the distraught father, guilt-ridden at having pushed that son beyond his capacities.

John Quincy's second son, John, expelled from Harvard for unruly behavior, was little better. As his father's private secretary and namesake, he assumed the President's power as his own. He was, in journalist Perley Poore's report, "obnoxious," once loudly insulting a leading Washington editor at a White House reception. In the Capitol rotunda a week later, he encountered the editor, who, as Poore recounted it, "pulled his nose and slapped his face. A scuffle ensued, but they were quickly parted." It was hardly the most effective way to win friends for the President in a hostile Congress. This Adams son also died an alcoholic, at thirty-one. The third son, Charles Francis, redeemed the family name—he served in Congress, was twice nominated for President, and was a lustrous ambassador to Great Britain, like his father and grandfather before him.

Like John Quincy Adams, John Tyler had made his son and namesake, John, his private secretary, and like the young John Adams, the young John Tyler took to drink. His father gave him no special privilege: he fired him. George W. Bush might have been another on this depressing list but for one thing—his own discipline. He recognized—prompted by an ultimatum from Laura—that he had become an alcoholic and took charge of his life before his father reached the White House.

The worst of this lot was Payne Todd, Dolley Madison's son from her first marriage. He was only three when Dolley married the distinguished James Madison, who would be the only father Payne ever knew. As a college student the boy showed signs of waywardness. Trying to settle him down, Madison sent Payne to Russia with the delegation to a peace conference, which failed to settle either the War of 1812 or Payne. In his long stay abroad, he acquired an incurable taste for high living and high debt, and in maturity he bankrupted Madison's Virginia plantation, filched valuable presidential papers, and swindled his mother. Payne Todd, the wastrel who had lived in the White House and moved in royal circles abroad, was ordered to debtor's prison—not once but twice. "My poor son," the usually wise Dolley lamented from behind her mother-tinted glasses. "Forgive him his eccentricities. His heart is good, and he means no harm."

The catalog of presidential sons plagued by alcohol and scandal includes at least nine in the nineteenth century but only one well-documented example in the twentieth century—Theodore Roosevelt's Kermit, an army major in World War II, an alcoholic who shot himself in 1943. Why did the syndrome virtually disappear in the twentieth century? Was the burden of the father's name greater for those who worked for their fathers? Greater because they were under closer scrutiny in a more open White House? Was a weakness worsened by constant obeisance to the father, living with him in the White House and serving as his right-hand man, rather than pursuing independent lives as the sons do today? Members of a President's family are now barred by law from being on the White House payroll; in view of the record, this seems a wise constraint.

OVER THE YEARS presidential daughters have caused their parents little more than a passing headache and not one recorded instance of alcoholism. The simple explanation may be that even though they will always be known by their father's name—Caroline has always been Kennedy; Chelsea will always be Clinton—they are not burdened with the same expectations. The public wants to see loving, supportive daughters who serve when needed and stay out of the headlines. The daughters are probably aided by the very things they resist: White House reins are held tighter on them, and they receive more public attention. And there's nothing like the spotlight to keep First Daughters walking the chalk line. The Bush twins learned that the hard way.

Surely the most welcome advice ever offered a White House daughter came from the most celebrated of that exclusive company, a daughter who delighted in not walking the chalk line—Alice Roosevelt Longworth. Almost seventy years after her own giddy White House years, "Princess Alice" wrote a note to Susan Ford as she entered the President's House: "Have a helluva time!" Susan enthusiastically followed the advice.

four

Love in the Fishbowl

———◦◦◦———

THEY CALL IT a fishbowl, with good reason: White House families must swim in full view, not only of the ever-present staff but of the ever-present world. Occasionally they may dodge scrutiny by hiding amid the plastic seaweed or lingering in the cute little castle. All to no avail. Long ago Chief Usher Howell G. Crim, who managed the White House through a series of administrations from Franklin Roosevelt's first term in 1933 to President Eisenhower's in 1957, expanded the metaphor: "The goldfish bowl is made out of magnifying glass." And never is spectator curiosity as intense as when a President is swimming out of the castle.

It was not until the tenth President, John Tyler, lost his wife in late 1842 that a chief executive's private life dominated the chatter over Washington teacups. His once-beautiful wife, Letitia Christian of Virginia's plantation society, had arrived at the White House partially paralyzed and confined to her room. The Tylers' daughter-in-law Priscilla took on the duties of First Lady, coached by the indomitable Dolley Madison. Letitia made only one social appearance in the White House, attending their daughter's wedding, on January 31, 1842, in the East Room—the same salon in which the First Lady's own funeral would be held seven months later.

Within a very short time, Washington was whispering about the recently widowed Tyler's obvious interest in Julia Gardiner, daughter of a New York senator and a young society belle thirty years his junior. The nattering ladies of capital society were shocked that the well-born Julia had lent her prestige to a rather crass advertisement for a New York department store. Though the illustration labeled its pretty model only as "the Rose of Long Island," it was clearly Julia holding a placard bearing a less-than-upscale endorsement: "I'll purchase at Bogert and Mecamly's. . . . Their goods are beautiful and astonishingly cheap."

Despite that venture and Julia's mother's dim view of him as a son-in-law, the President wooed and won. Paranoid that the venue of their wedding in New York might leak to the press, he had the staff of his hotel held under lock and key until after the ceremony. In only eight months as First Lady (Tyler was not nominated for a second term), the glamorous Julia ignited the White House and won over skeptical Washington, which will forgive almost anything if the parties are good.

Tyler holds the distinction of having two First Ladies in less than one full term in the White House, considerably more to be desired than the sole achievement of his predecessor—William Henry Harrison gained his footnote in history as the first President to die in office, the first to lie in state in the White House.

IN 1864 GROVER Cleveland, a rising lawyer in Buffalo, was faced with a problem that was not easy for a young bachelor of twenty-seven. What sort of gift would be suitable for his law partner's newborn daughter? After giving the matter considerable thought, he settled on the ideal present for infant Frances Folsom—a baby carriage.

Twenty-one years later, as President of the United States—still a bachelor—he gave even more serious thought to "Frank," then a beautiful Wells College senior: he asked her to marry him. Throughout the years between her babyhood and blossoming there had been a close familial tie; as administrator of her late father's estate, Cleveland had kept watch over her as his ward. Close watch.

It was the ultimate May-December romance; he was a portly forty-nine, she a lissome twenty-one. While Frances was still a college student, the President, the soul of propriety, asked her mother's permission to write to her, discreetly. This was followed by a number of equally discreet visits by Frances and her mother to the White House, ostensibly as the guest of the President's sister, who was his hostess. Once again propriety required Cleveland to wait for two months after Frances's graduation before proposing—by letter. She accepted, apparently not put off by his acknowledged illegitimate child, an awkward fact that surfaced during the campaign of 1884. (Republicans had taunted him: "Ma, Ma, where's my Pa? Gone to the White House, ha, ha, ha!") Further discretion decreed that Frank and her mother take a trip to Europe, where she could assemble a trousseau without being noticed.

Such punctiliousness duped both friends and press, who assumed that the President was a confirmed bachelor—in April 1886 the *New York Herald* dismissed any ruminations to the contrary as "incredulous." It was not the first or last time the press had such White House rumors wrong; the following month the engagement was announced.

To avoid the problem of a stampede when Frances disembarked in New York, a presidential confidant intercepted the ship in a customs-house tug and whisked Frances and her mother ashore and off to Washington, leaving the press gang high and dry at the dock. But the word was out and the reporters were in full cry.

Frances was tall and comely, with sparkling brown eyes under strong dark brows and a wasp waist that was the height of fashion. She instantly became the public's darling. For this first—and still the only—White House wedding of a President, the chief executive became his own social secretary, taking charge of every detail. He personally wrote each invitation to the twenty-eight invited guests:

> My dear Mr. ———,
>
> I am to be married on Wednesday evening at seven o'clock at the White House to Miss Folsom. It will be a very quiet affair and I will be extremely gratified at your attendance on the occasion,
>
> Yours sincerely,
> Grover Cleveland

It is unlikely that anyone declined *the President's* wedding.

On the night of the ceremony, a crowd gathered outside the Executive Mansion to watch the select few guests arrive. They strained to hear the music of the Marine Band, led by the celebrated John Philip Sousa, and craned to catch a glimpse of the President and his young bride. Ever-present journalist Perley Poore, who was allowed inside, savored the details: the couple descended the stairs together; the bride's ivory satin gown, a creation of sundry loops and folds and borders of orange blossoms, flowed into a four-yard train, overlaid with a five-yard veil of tulle. No one could miss her sparkling diamond necklace, the President's wedding gift. Poore duly described the bridegroom's "evening dress of black, with a small turned-down collar . . . and a white rose fastened to the lapel."

The President had trimmed the ceremony down to a brisk ten minutes, marked by a twenty-one-gun salute from the navy base and church bells pealing across the city. After the reception the couple left "amid a shower of rice and old slippers" to board a special train to Deer Park, Maryland, where the meddlesome press were waiting, staked out in choice spots with spyglasses and direct telephone lines to their editors. The couple quickly cut short their honeymoon for the relative privacy of the White House. It was the new First Lady's first lesson in the aggravation that comes with the title.

IN THE DARKENED East Room Woodrow Wilson paced back and forth, back and forth, beyond the comfort offered by his three daughters. The President was distraught over the death of his beloved wife, Ellen, after a blissful marriage of almost thirty years. "I can't bear it," he said, weeping. "I can't go on without her."

Yet he did, with remarkable haste. Less than eight months after Ellen's death in August 1914, Helen Bones, his cousin who had moved into the White House, introduced the President to her friend Edith Bolling Galt. At forty-three, Edith was a pretty and vivacious widow with a penchant for Paris gowns, a saucy lady who drove herself around Washington in her own electric automobile. The President, whose public austerity belied hot passions raging within, was instantly smitten, which quickly became plain to see—and just as quickly Washington buzzed with gossip about the heart worn on the President's sleeve so soon after his wife's death.

Edith herself was troubled by that, and told him so. In her memoirs she recounted his reply: "Yes, I know. But, little girl, in this place time is not measured by weeks, or months, or years, but by deep human experience." Edith could save him from a lifetime of loneliness, and in September 1915, she agreed to a secret engagement.

With his bid for reelection in 1916 already in motion, Wilson's advisers were dead set against the marriage so soon after his wife's death. James Kerney, publisher of the Trenton, New Jersey *Evening Times* and Wilson confidant, described the quandary: "It was a case of love at first sight. In his courtship of Mrs. Galt, Wilson displayed all the ardor and intense eagerness of a boy. Coming so soon after the death of Ellen Axon Wilson, it gave the Democratic politicians a great fright. But none in the Cabinet dared to speak to the President of their misgivings." When the President's private secretary, Joseph Tumulty, dared suggest delaying the marriage until after the election, he "never again enjoyed the favor of the family circle."

For privacy the courtship was conducted within the protective arms of the White House, chaperoned—to what extent?—by cousin Helen, but the eager President found a way around the constraints. Years later his Secret Service agent, Colonel Edmund Starling, revealed their trysts: "In order to be alone—as much alone as possible—with her, the President added a walk in Rock Creek Park to his afternoon automobile ride." The vast park, a swath through the heart of the city, is even today almost a mountain fastness in some areas, with paths well away from the eyes of motorists. Charged with keeping the President in sight every moment, Colonel Starling was in an awkward spot: "I wanted to look away. I wanted to let a tree get between me and the two of them. But I couldn't." In the line of duty, the agent witnessed the President in love: "He was an ardent lover. . . . He talked, gesticulated, laughed, boldly held her hand. It was hard to believe he was fifty-eight years old. Every now and then they would glance back at me, he with the embarrassed half-smile of a man who wishes you would go away; she with the frank laughter of a woman who is enjoying the predicament of both men. She was having a wonderful time."

To escape the glare of the White House, the President rented a summer place in New Hampshire—and Mrs. Galt was there. Colonel Starling could see that "his eyes sparkled as he cast frequent glances in her direction. He was plainly thrilled by it all." The scholarly President joked, danced, and would break into his favorite song, "Oh, you beautiful doll, you great big beautiful doll!" Away from the public eye, the President whom history limns as austere and long-faced revealed the inner man. Forty years later his daughter Eleanor Wilson McAdoo challenged the "myth" that her father was "a very solemn man, an intellectual snob," insisting that "in the bosom of his family, it took little urging to get him to go into his vaudeville act." James Kerney added another dimension: "When he set out to be, Wilson was especially fascinating to women. He did not care for male company, but he enjoyed the admiration and adulation of women."

The President and his merry widow were married on December 18, 1915, in her Washington home, making him the third President to wed while in office. Edith was then with him constantly—on the golf course, the campaign trail, and official trips abroad (the first for a President in office). Observing them daily, Chief Usher Ike Hoover noted, "The President and his wife moved together in perfect accord, each much the better for having joined the other." As the election neared, Starling noted that "Wilson took little interest in the campaign. He was on his second honeymoon and usually made but a single speech a week, each Saturday afternoon." Now, that's a President in love.

Edith proved to be much more than the light in his darkness. Twenty-two days into his exhausting national tour in 1919, his crusade to sell America on his vision for a League of Nations, Wilson was felled by a stroke. From that moment Edith and members of her family, in great secrecy, assumed the role of virtual President, hiding his condition, keeping a suspicious cabinet and Congress at bay, carefully staging his occasional brief appearances. Though she handled the charade with Machiavellian skill for a year and a half, it was a very wrong—and surely unconstitutional—course of action. In *Edith and Wilson,* historian Phyllis Lee Levin sums up Edith as "a dissembling and unworthy figure in the history of the American presidency."

WHILE THE EXECUTIVE Mansion is always considered the domain of the First Lady, the four widowed nineteenth-century Presidents—Thomas Jefferson, Andrew Jackson, Martin Van Buren, and Chester Alan Arthur—set the ladies of Washington to speculating and scheming. Two of the three Presidents whose wives died during their presidency (Tyler and Wilson) quickly—very quickly—took on new ones. The third, Benjamin Harrison, lost his beloved Caroline only days before he was defeated for reelection—he was not sorry to be rejected, for he could not have borne the White House without Carrie. Jefferson's young wife had died nineteen years before his election, but surrounded by his daughters and grandchildren—and his books—he was not lonely in the White House.

The absence of wives led to a variety of official stand-ins, mostly young and quite delighted daughters, daughters-in-law, and nieces. But from its first sketched outlines, the stately residence was meant to be shared, and the burdens of office must have occasionally cloaked those single Presidents in personal isolation—which may account for both Tyler and Wilson remarrying so soon as to cause scandal. Benjamin Harrison waited four years before marrying his wife's niece, who had been her White House social secretary.

The two bachelor Presidents followed very different paths. In 1857 lifelong bachelor James Buchanan, with his impeccable credentials as former congressman, senator, diplomat, and secretary of state—not to mention his wealth—was catnip to the ladies of Washington. When he proved impervious to their scheming (he was, after all, almost sixty-six), it was bruited about that he had a longtime mistress, but she remained a will-o'-the-wisp. Middle-aged bachelor Grover Cleveland moved into the White House with secret intentions that ended his single status and ultimately brought him five children and contentment.

MARGARET BAYARD SMITH, the legendary chronicler of *The First Forty Years of Washington Society,* confessed her own reaction when, in 1801, a tall, rangy gentleman, a stranger, knocked at her door, inquiring about some printing work (her husband published the leading newspaper). "He turned to me . . . an expression of benevolence and with a manner and voice almost femininely soft and gentle," she gushed. "I know not how it was, but there was something in his manner, his countenance and voice that at once unlocked my heart." Then she learned who the caller was: "I felt my cheeks burn and my heart throb." From that moment, Mrs. Smith, formerly a strong Federalist and a John Adams supporter, was a captive of Thomas Jefferson. Not until John F. Kennedy did another President have such an effect on tough-minded women journalists. His charm and erudition made Jefferson a perfect target for Washington matchmakers, but he proved impervious to their machinations. There were occasional rumors about his close friendship with Dolley Madison, which were dismissed as malice circulated by his enemies. Too scandalous to be mentioned in polite society was the published charge that Jefferson had fathered children with his house slave, Sally Hemings, which is now widely accepted as true.

No President was ever so devastated by malicious rumors as Andrew Jackson, who would always maintain that slander against his beloved wife, Rachel, was the cause of her death—only ten weeks before he entered the White House. She had never wanted to leave their Tennessee plantation, never wanted to return to Washington, where she had lived when her Andy was in the Senate. When Jackson won the presidency in 1828, Rachel was asked the usual how-do-you-feel question. With candor rare in politics, she replied, "Well, for Mr. Jackson's sake I am glad. For my own part, I never wished it." To her friends she confided, "I'd rather be a doorkeeper in the House of the Lord than to live in that palace."

Her experience of Washington led her to deplore the capital's preoccupation with "carding and running to parties," which suggests that Rachel would not have enjoyed the role of First Lady. Not that she was a prude. A well-born Virginian, Rachel, though far from pretty, was a jolly lady who enjoyed dances and the races when her husband's Thoroughbreds were running. Her genial hospitality at the Hermitage, their showplace estate near Nashville, would have prepared her to supervise a well-run Executive Mansion, and her calming presence would surely have benefited the turbulent Jackson White House.

Theirs was an enduring love that never lessened through the rootless life of the military and wretched days of political slander. The fiery general, a hero of the War of 1812, had won the popular vote in 1824, only to lose in the House of Representatives after John Quincy Adams entered into an unseemly deal with Henry Clay. Jackson, a Carolina native who had gone "west" to East Tennessee, was the first of the frontier Presidents.

In Tennessee he married Rachel Donelson Robards, an even-tempered, deeply religious Presbyterian, divorced from a harsh and jealous husband, Lewis Robards. Or so they thought. Robards had, in fact, petitioned for divorce but never completed the process. Andrew and Rachel, after two years of marriage, were stunned when Robards reemerged with proof that the divorce had not been finalized and then divorced Rachel on grounds of adultery.

The ugly action would be a political millstone and the source of malice that would trigger Jackson's volatile temper—in one duel he killed a man who had slurred his Rachel. During the brutal campaign of 1828 Rachel wrote, in her eccentric spelling and grammar, to an old friend: "The enemys of the Genls have dipt their arrows in wormwood and gall and sped them at me Almighty God was there every aney thing to equal it. . . . Theay have Disquieted one that theay had no rite to do theay have offended God and man."

The worst, it is told, came when Rachel, shopping in Nashville for a White House wardrobe, overheard women assailing her character, castigating her as an adulterer, ridiculing her as a First Lady. Rachel, hysterical, inconsolable, told her niece Emily, "I'll never forget it. I will not go to Washington, but stay here as often before in Mr. Jackson's absences." Her health, already impaired, failed under the cruel assault, and on Christmas Eve she was buried, clothed in the white satin gown she had chosen for the inaugural ball. In death the wife who would never become First Lady was honored—all across Tennessee church bells tolled the loss; at their plantation home ten thousand came to mourn her passing.

Jackson would always believe that Rachel had died of a broken heart, for which he blamed the vilification spread by agents of the Adams White House. "May God Almighty forgive her murderers, as I know she forgave them," he declared at her funeral. "I never can. . . . Those vile wretches who have slandered her must look to God for mercy." Throughout his presidency the old general wore a miniature of "my dearest heart" around his neck and each night placed it at his bedside. There would never be another woman in Andrew Jackson's life.

CHESTER ALAN ARTHUR was a widower of twenty months when President Garfield's death propelled him into the White House in 1881, igniting a flurry of matchmaking along with rumors. His resistance to the matchmakers' schemes, coupled with his secretive late-night suppers, led to talk that the wealthy lawyer and political boss had a mistress. For a time, speculation centered on Frances Willard, a temperance leader, an odd notion since Arthur, a New York sophisticate, could often be found by the punch bowl. Then talk turned to the daughter of a Supreme Court justice. That there was not a trace of evidence for either rumor didn't staunch the flow.

Apparently quite content to have his married sister as his White House hostess, with her two young daughters as company for his motherless nine-year-old Nell, the dapper President pursued a lively social life, driving about town in his smart deep-green landau drawn by a pair of fine mahogany bays, dining at all the best homes, vacationing in fashionable Newport, and was roundly criticized for his frivolity.

The gossips were unaware that the President kept a portrait of his entrancing Ellen at his bedside with fresh flowers every day, or that the new stained-glass window in St. John's Church, "the Presidents' Church" on Lafayette Square, was his tribute to her memory. The window, still one of the old church's treasures, was installed on the south wall; from the White House the President could look across the park and see the church lights glowing through its jewel colors.

Had Washington's ladies been allowed to vote in 1876, Democrat Samuel Tilden, governor of New York, would surely have carried the District of Columbia. Shortly before the election the Albany (N.Y.) *Knickerbocker* reported that the wealthy bachelor's "greatest wish, after his election as President, is to secure an accomplished and good woman as a wife . . . who will be every way worthy of the high honor of being the partner of the Chief Magistrate of the nation, and who would officiate at the White House receptions in a dignified and womanly manner."

Every unmarried woman in the capital was confident that the job description fit her as snugly as the glass slipper fit Cinderella. Today's reality television could not script such a scramble—the ultimate bachelor choosing from a line-up of finalists in the Executive Mansion. But in real reality, when the deadlocked election was at last settled in favor of Rutherford Hayes, amid bitter controversy, the bachelor-governor called off the search.

BIG, BEEFY PRESIDENT Taft was a tender soul where his wife was involved. To others, Nellie was an opinionated pinchpenny and generally difficult woman, but his eyes saw a different First Lady, a hostess who held her own in following the much acclaimed Roosevelt years, a confident lady who drove herself about town in her own small electric "landaulette," a virtual commanding officer in overseeing her many changes in the Executive Mansion.

On their way to an official function one afternoon, Nellie suddenly slumped, immobilized by a stroke. "The President looked like a great stricken animal," Major Archibald Butt wrote to his mother. "I have never seen greater suffering or pain shown on a man's face." Taft's devotion to Nellie led to an unseemly consequence. Strolling with Major Butt, the two stopped to admire the showcase flower beds at the Department of Agriculture, and the President picked a few roses for Nellie, recuperating from her stroke. It was an act of love—and an act of vandalism that brought an old watchman running, yelling, "You are under arrest!" Suddenly he recognized the thief he was collaring. "I have to obey my orders whether it is the President or not," he stammered. "Right you are," said Taft genially.

"Do you want me to go to the station house with you?" The watchman gulped, the matter was dropped, and back in the White House Nellie was cheered by flowers filched just for her.

Love of a different kind was gently expressed by President McKinley for his wife, Ida. Together they had suffered the loss of their only children, two baby daughters, and later he gently coped with her epilepsy. Most of the time she stayed in her room, but she gamely attended social functions, always with her husband close by her side.

Countess Cassini, a White House regular, described a delicate scene at an official dinner: "Mrs. McKinley, looking fragile and drawn . . . suddenly, making no sound, stiffens in her chair and begins to quiver violently. Calmly, the President throws his pocket handkerchief over her face, rises, lifts her gently in his arms and carries her from the room without a word. In this moment he is great. In a moment he is back and resumes his place as if nothing has happened." Knowing how much Ida, a rather petulant and irritable woman, enjoyed the role of First Lady, the deeply devoted McKinley encouraged her to keep up her social engagements, even as her condition deteriorated. At the moment he was fatally shot by an anarchist, his first thought was of Ida: "My wife," he murmured urgently to an aide, "be careful how you tell her."

The dying words of two other Presidents declared eternal love for their wives. "I love you, Sarah, for all eternity," James Polk declared. Rutherford Hayes, whose wife had died more than three years earlier, whispered, "I know that I am going where Lucy is."

THERE HAVE BEEN moments that would have been best left within the miniature castle in the White House fishbowl, incidents that generate livelier discussion about the President's private life than his policies. Though it may be only rumor or exaggeration, White House gossip spices up the afternoon tea and can be the olive in the martini.

In 1941 Chief Usher Crim, briefing his new assistant on President Franklin Roosevelt's staff, gave studied emphasis to one particular title. "Missy LeHand is the President's *personal* secretary," he told J. B. West. "She probably sees more of the President than Mrs. Roosevelt does. She acts as his hostess when Mrs. Roosevelt isn't here." Missy (Marguerite) was one of three Roosevelt aides who lived in the White House, quite at home in a pleasant third-floor suite.

"In contrast to Mrs. Roosevelt's close relationship with friends and her husband's with his staff," West observed years later, "we never saw Eleanor and Franklin Roosevelt in the same room alone together. They had the most separate relationship I have ever seen between man and wife. And the most equal."

White House seamstress Lillian Parks, writing her own story, detailed the special relationship between President and secretary: "She was the one who shared his private jokes, the one who first learned of his ideas, and the one who applauded them without reservation. She was always there. . . . She kept *his* hours, eating dinner with him in his office when the First Lady was not around. She even went swimming with him to keep him company though she hated the water. Backstairs we used to wonder if 'Mrs. R' wasn't a little jealous of Missy LeHand, but she never seemed to be."

In the backstairs view, conveyed by Parks, "Missy gave him the companionship, the rapt attention, the ego-building boost that men sometimes find in their wives. 'Mrs. R' was not the kind of woman who would give blind praise or blanket approval." Parks insisted that "there was definitely no question of anything improper. It was a spiritual attachment, although Missy built her whole life around him and never married." Four of the five Roosevelt children shared that view; son Elliott held the opinion that the attachment was quite different.

Eleanor later wrote with cool detachment, "Franklin might have been happier with a wife who was completely uncritical. That I was never able to be, and he had to find it in other people." Perhaps by then she no longer cared. "Nevertheless," she pointed out, "I think I sometimes acted as a spur, even though the spurring was not always wanted or welcome." Yet the sound of hurt was there: "I was one of those who served his purposes." The White House gave Eleanor a platform, a life of her own, a place in history. Had she been a "Missy" wife, Mrs. Franklin D. Roosevelt would not have become a world humanitarian, the subject of many biographies, the recipient of great honors, and—seven decades later—still admired as a trailblazing feminist.

After all of the principals were long gone, Chief Usher West revealed a story featuring a different lady: "Quite often, but only when Mrs. Roosevelt was out of town, the President invited his friend Mrs. Lucy Mercer Rutherfurd to the White House. An attractive woman in her forties, she'd arrive at the front door, the north entrance. We'd watch her hurry up the steps, to be escorted by an usher to the second floor. The butler would serve tea, close the door, and leave the President and Mrs. Rutherfurd alone. After about an hour's time, the President rang for the doorman to escort her back to her car." And now and then Mrs. Rutherfurd would be the guest at dinner with the President and his daughter, Anna Boettiger, but always without Eleanor.

FDR's frequent drives into the Virginia countryside caused knowing amusement among the staff. Driving along a wooded lane that he had specified, the President would "notice" a lady walking along the road. "Let's ask her if she needs a ride," he would say to his driver. It seems that she did. The fourth time these same circumstances occurred, the Secret Service agent noted that "they always took the long way around." In response to his agent's curiosity, the President only laughed.

This was not merely an attractive lady "needing a ride" on a back road in Virginia. More than twenty-five years earlier, when Franklin was assistant secretary of the navy, the charming Lucy Mercer had been Eleanor's social secretary. Franklin fell in love with her and wanted to marry her. Eleanor, the submissive wife in those days, agreed to a divorce if that was what he wanted. Then his indomitable mother, Sara Delano Roosevelt, declared that if Franklin, the father of five, divorced Eleanor—thus ruining his promising career—she would cut him off from the family fortune and his beloved Hyde Park home. Sara's unblinking command to the son who was her life could be said to have affected the course of American political history. Son bowed to mother's will. He gave up Lucy; he gained the White House and became one of the nation's greatest leaders.

Yet years after that turning point in Franklin's life, that same Lucy, then properly married and still charming, would meet him on a country lane and join him for dinner at the White House when Eleanor was away. And on that fatal day at his personal retreat in Warm Springs, Georgia, it was Lucy—still his love?—who was at Franklin Roosevelt's side when he was struck by a fatal cerebral hemorrhage. And once again Eleanor would feel betrayed.

Perhaps the most scandalous gossip to plague a President swirled around Warren Harding even before he gained that office. Washington had been whispering about the married senator's liaison with Nan Britton, a pretty blonde thirty years his junior, and the daughter who was born in 1919, the year before he was elected President. It was a devastating rumor, but Harding made no reply—he could not, because the story was true. (Chief Usher Ike Hoover called Harding "a sporting ladies' man.") Yet, though the facts were widely known, it did not derail his nomination as the Republican candidate for President in 1920. To his credit, Harding quietly sent money by his broker every month to support the child he refused to see.

Later, Nan spelled it all out in her kiss-and-tell book, *The President's Daughter*, which was made into a film, *Children of No Importance*, in 1928. She told of trysts in a small closet off the anteroom to the Oval Office (possibly the venue that figured in the news in 1997), where "we could share kisses in safety. We repaired there many times in the course of my visits to the White House." The daughter, Elizabeth Ann Blaesing, who died in 2006 at age eighty-six, is a rarity in today's cash-in society—she clung to her privacy in Oregon, her longtime home, and never spoke publicly about her White House connection.

But there was another woman, less known but more truly loved, in Harding's life: Carrie Phillips, wife of the owner of a large department store in Marion, Ohio, Harding's hometown. Over a span of fifteen years, Carrie and Harding, who was climbing the political ladder from newspaper publisher to state senator, lieutenant governor, and then U.S. senator, shared a secret liaison. It began in 1905 and continued until early 1920, with a hiatus during the prewar period from 1908 to 1912, when Carrie and her daughter lived in Europe—mostly in Germany, where she became sympathetic to the Kaiser.

When Harding decided to run for President, the affair ended, though they continued a friendship. According to reliable sources, Harding's key supporters (led by Will Hays, who would later become Hollywood's powerful censor), fearing the scandal might surface and jeopardize his chance for the White House, stepped in with preventive action. They paid Carrie and her husband twenty thousand dollars plus a monthly stipend to take a very long trip around the world, which they did, remaining abroad until after Harding's death.

Letters between Harding and Carrie continued and are under seal at the Library of Congress until 2014. Harding's biographer, Francis Russell, who apparently saw copies, described some of them as "shocking . . . more so because they were written by the President of the United States than the tumescence of their content." In *The Shadow of Blooming Grove,* Russell wrote sympathetically, "Carrie Phillips was clearly the love of his life, and he was more loving than loved." But, then, Nan Britton was still around.

When the inconversable Vice President Calvin Coolidge succeeded Harding, Washington gossips put away any hope of new material. But who would have thought that deep within the flinty
Cal—famously described as having been "weaned on a pickle"—
burned a jealous heart. The single exception to his social frostbite
was his wife, Grace, who, even in the little-noticed role as wife of
the Vice President, was the most popular woman in Washington.

As a husband, Coolidge was a martinet, angry if she was a
minute late coming home, not allowing her to drive a car or fly in a
plane, even with Charles Lindbergh at the controls. When, as First
Lady, she was learning to ride horseback—without consulting
him—his response was a withering "I think you will get along on
this job fully as well if you do not try anything new," an approach
that pretty well summed up his presidency. In 1928 Grace had no
inkling that he would not seek reelection, a decision affecting her life
as much as his, until she heard his characteristically laconic public
statement: "I do not choose to run."

During a summer vacation in the Black Hills of South Dakota,
Cal triggered a White House scandal when Grace went for a walk,
accompanied as usual by a Secret Service agent—a personable fellow, new on the job. As lunchtime came with no Grace in sight, the
President grew worried. Time passed; he paced the porch of their
lodge; a search party was sent out.

An hour and a quarter later the First Lady showed up, carrying
wildflowers she had picked, explaining that she and the agent had
gotten lost. The President was a thundercloud. For days he was
sullen, silent, locked in his jealousy. Reporters covering him witnessed the episode, and rumors of divorce followed. On their return
to the White House, social secretary Mary Randolph, an early spin
doctor, dampened the gossip by having them take walks together for
all—especially photographers—to see. Grace, according to Miss
Randolph, was "the sunshine and joy of his life—his rest when tired,
his solace in time of trouble." In his stony, introverted way Cal Coo-

lidge adored his Grace. She understood and accepted all that came with it.

THE WHITE HOUSE, with its prestige and trove of emoluments, can anneal the fault lines of faded marriages, bringing partners closer together in their life's great adventure—or its relentless social and psychological demands can unbalance the familiar equation of comfortable unions. Both the Eisenhowers and the Kennedys were reliably said to have become closer; Louisa and John Quincy Adams, arriving in 1825, did not; others, notably the Franklin Roosevelts, comfortably accommodated their separate interests.

In one way or another, "the great white jail" warps what passes for normal life—an effect that is not entirely without an upside. Its very restrictions enforce more togetherness, encouraging greater dependence on each other for comfort and support. Lady Bird Johnson remembers saying to Lyndon, "It's just you and me. Nobody can do it for us. It's up to us to make it a success." Some couples come into the White House as perfect partners—Rosalynn and Jimmy Carter shared every concern, whether family or international; George and Barbara Bush had an unwavering bond that began the moment they met as teenagers at a country-club dance; for Nancy and Ronald Reagan the White House was the closing sequence of their Hollywood movie of a life.

Some lucky presidents are like Eisenhower, who was rediscovering his buoyant Mamie after being separated from her for so many years during World War II. Celebrating their fortieth wedding anniversary at a dinner with friends, Ike—a general not given to sentiment—lifted his glass to his wife; being married to Mamie, he declared, meant ever more enjoyment as the years passed. Mamie, looking younger after eight years in the White House than she did upon entering it, glowed; the White House had more than made up for the time they had lost.

During the John Quincy Adams term, the ladies of the White House social circle (who were often more at home there than the First Lady who dutifully invited them to tea) felt it necessary—their duty, they would say—to keep close tabs on the personal atmosphere of the President's House. Otherwise, how would people know the inside stories? What their sharp eyes observed then added tang to their after-tea gossip. So they took notice when John Quincy Adams, "a silent animal, a man of austere and forbidding manner" by his own account, glowered at Louisa, and the grapevine was busy when Louisa often responded sharply. In the White House fishbowl a dead love is as difficult to conceal as a budding romance.

Louisa evinced no interest in politics or her position as First Lady, which is odd given her role as a leading Washington hostess when her husband was secretary of state with his sights on the presidency. Her charm cloaked his lack thereof; she was an accomplished musician, spoke perfect French, and was the real diplomat in the family. To advance her husband's ambition to win the White House—to fulfill his father's expectations?—she had made every wise political move, entertaining brilliantly, cultivating the powers on Capitol Hill, lobbying the influential, even traveling to meet nearby state leaders. That she candidly called her efforts "my campaign" implies a lively interest in a shared goal.

Yet once the prize was won, Louisa lost interest—or was pushed aside. She commented tartly in her journal, referring to someone seeking her help for a government position, "They are woefully mistaken if they imagine I had the least influence. I believe it would be sufficient for me to utter the name of any person . . . to have them excluded all together." That suggests an escalation in their mutual accusations, and denials, of personal ambition. In the Executive Mansion Louisa retreated to her room as a semi-invalid, stricken with melancholia, hysteria, and fainting spells. Significantly, she revived quickly after their one term, freed of the insidious effect of the White House.

Further clues to her White House discontent can be found in her diary: "I have since the first year of my marriage entered upon my great honours with tears. . . . I have nothing to do with the disposal of affairs and have never but once been consulted." Clearly John Quincy shared nothing of substance with the intelligent and admired wife who had done so much for his career. Neglected and resentful, she retreated into herself.

Their youngest son, Charles Francis, after graduating from Harvard, moved into the White House to study law with his father. In his diary he summed up his parents' fractious relationship with a dismal observation: "A more pitiable set I do not think I know."

How different it was with John Quincy's father and mother. During their many years apart John Adams and Abigail, his intellectual equal, constantly exchanged letters, sharing endearments, family news, and political opinions. She sent him off to Paris with 1,500 sheets of paper and plenty of ink—along with five bushels of corn, six chickens, two sheep, and a ten-gallon keg of rum. (Their correspondence takes up 608 reels of microfilm in the Adams repository.)

Belying his formidable manner, Adams wrote to the wife he privately called "Miss Adorable" with unmuzzled passion. In 1797, only two weeks after his inauguration in Philadelphia, the nation's second President dispatched a plea to his wife, who had expected to manage their Massachusetts farm through the summer: "I can't live without you till October." A month later he demanded, "I pray you come on immediately. I will not live in this state of separation. Leave the place to my brother—to anybody or nobody. I care nothing about it—but you I must and will have. Leave the place to the mercy of the Winds." What wife could resist such longing? Not Abigail. In a flash she was on her way to Philadelphia. History does not portray John Adams as a passionate man, but to Abigail he was soul mate.

Before the era of telephone, e-mail, and easy visits, letters were windows opening on heart, soul, and mind. Among the most tender presidential letters are those between James and Lucretia Garfield—

"Jim" and "Crete." They give us an inkling of the atmosphere of the Garfield White House, which tragically ended with his assassination after only four months. In the impersonal quiet of the Library of Congress manuscripts room, it seems almost an intrusion to read their letters, overflowing with love and longing for each other during his service as a major general in the Civil War and in Congress. He addressed her "My Own Darling," "Dearest Love," "Dearest Darling"; she responded "My Darling Precious," "My Dearest Life." The White House had promised happiness for the Garfields, their five children, and his much admired mother, who was proud to be the first mother to attend a son's inauguration.

Their letters conveyed a different tone, more personal, less intellectual, than the famous correspondence between John and Abigail. The two sets of letters reflect the cultural distance between eighteenth-century Puritan New England and the less restrained new West of the post–Civil War era. But in their intimate exchanges, both couples reflect the level of prevailing order giving way to a more urgent level of ardor.

AT FIRST GLANCE, when you look at pictures of Bess and Harry Truman together, you might not think romance. But what you see there is true and abiding love—the love that first struck six-year-old Harry's heart when his eyes fell on this golden-haired girl in Sunday school in Independence, Missouri. It took him twenty-nine years to win her, but for him there was never another girl, and never would be. During his ten years in the Senate, Bess was his partner, running his office (for which she was paid a salary, a practice that is no longer allowed) and discussing all issues with him.

She had not wanted him to run as FDR's Vice President—neither had he—and she recoiled at the very thought of the White House. Worse yet, she found herself sealed out of their long working partnership—as chief executive, her husband was now surrounded by advisers who knew more about the presidency than he did; he was working long hours grappling with crucial decisions on the war, matters too highly classified for him to share even with Bess. First among them was the overwhelming revelation of the weapon even he as Vice President had known nothing about—the atomic bomb. He looked history in the eye and loosed the weapon of mass destruction over one Japanese city, Hiroshima, and then another, Nagasaki. For the first time in their long, close life together, Bess, his trusted partner, was kept in the dark, left out of the defining action of his life. Harry woke one night to find his stoic Bess sobbing into her pillow, a victim of the White House.

Their daughter, Margaret, watching the change the White House had imposed on her mother, later described how deeply it affected Bess. "She was forced to face a very unpleasant fact," Margaret wrote in her intimate biography of her mother. "She had become a spectator rather than a partner in Harry Truman's presidency. That made her very, very angry." As Bess "felt more and more superfluous," her reaction became "a smoldering anger that was tantamount to an emotional separation." Her despondency—"the White House blues," in Margaret's phrase—was surely a reason for her many long stays in their old home in Independence: escape.

It boiled over during the first Christmas of his presidency, when Harry, embroiled in critical negotiations with the Russians, had to remain in the White House until Christmas Day, while Bess and Margaret were already in Independence. Racing to join them, he ordered his plane to fly through weather so fierce that commercial airliners were grounded—only to be confronted by a glowering Bess: "So you've finally arrived. As far as I'm concerned you might as well have stayed in Washington."

Had that not been written by Margaret herself, it would be hard to believe a First Lady could be so harsh to a President coping with a grave world problem. In that one instance Harry, back in the

White House, popped off an angry letter to Bess—unthinkable for him—but the following morning he called Margaret and instructed her to retrieve his letter from the post office and burn it. Perhaps her little bonfire saved a marriage endangered by the White House.

The President took care to renew the partnership. In one month in 1947, when she was yet again in Independence, he wrote to her twenty-two times; over a span of fifty years, Harry wrote her, by biographer David McCullough's count, more than, 1,300 letters. On their wedding anniversary in 1948, he poured out his heart to her: "Twenty-nine years. It seems like twenty-nine days. You are still on that pedestal where I placed you that day in Sunday School in 1890. What an old fool I am." Oh, no—Harry was a man still in love with his wife, and Bess would have read it that way.

Years later Chief Usher West provided a delightful glimpse into the lasting romance between Bess and Harry, the story of Bess's first night back in the White House after spending another long summer in Independence:

"After a light dinner in the President's library, they sent the maids downstairs. The next morning, I was in Mrs. Truman's study at nine, as usual. In a rather small, uncomfortable voice, she said, 'Mr. West, we have a little problem.' She cleared her throat, demurely. 'It's the President's bed. Do you think you can get it fixed today?' 'Why certainly,' I said. 'What's the matter?' 'Two of the slats broke during the night.' 'I'll see that it has all new slats put in,' I said hurriedly. 'It's an old antique bed anyway, and if he'd like a newer one . . .' 'Oh, no,' she said. 'This one is just fine.' But the Trumans certainly aren't antiques, I thought to myself. The President's wife was blushing like a young bride."

Truman's flood of letters to Bess shine a light into both his presidency and his heart. How regrettable that the letters of some Presidents were purposely destroyed. Edith Roosevelt burned those of the articulate, tempestuous Teddy—possibly she felt they were too personal; Florence Harding destroyed all of Warren's—possibly she feared they were incriminating.

five

Whose Life Is It Anyway?

——➤·◦·◄——

"CAN YOU IMAGINE losing your anonymity at thirty-one?" Jacqueline Kennedy asked plaintively as she looked ahead to the White House. She was obsessed with privacy and control, and saw both at odds with the President's House that would be hers to define—fair warning of bumpy days ahead. She was right in foreseeing a public future, but she could not have imagined how public; in less than three years Jackie's sensuous face was recognizable the world over, and for the rest of her life she would be stared at, like the sighting of a rare endangered species. Her ever-present dark glasses had nothing to do with sun glare; they were her window blinds, blocking interaction.

For White House daughters and sons the spotlight can be even more disturbing. They arrive unaccustomed to public attention. Even the children of a Vice President are rarely troubled by the intrusion of the media and a gawking public; their official home, a large Victorian house, is both protected and well screened amid the expansive wooded grounds of the Naval Observatory on Washington's Embassy Row. White House children live in a much more constricted setting, where a hungry pack of photographers, shutters at the ready, is based in the White House press room.

At the heart of the commingling of love and hate that a President's family holds for the White House is the overnight mutation from private individuals to public figures. Even Martha Washington, who lived on a pedestal, though never in the yet-uncompleted White House, said with a sigh, "I think I am more like a state prisoner than anything else. There [are] certain bounds set for me which I must not depart from. I would much rather be at home." Once she is ensconced in the White House, every First Lady experiences the shock of suddenly becoming an object of unquenchable interest about what she does and doesn't do, her children, her clothes, her plans, her views. Overnight she is expected to give articulate answers about many subjects she feels are too personal, others that she has not yet had time to think about at all.

She learns what it's like to be stared at as though she were a mannequin in the Smithsonian's First Ladies exhibit, to be critiqued, described, quoted, misquoted. She discovers that, like it or not, by virtue of her new title she is no longer merely the wife of the man at the podium, she is news. At some point a new First Lady also learns that family secrets find their way into the press through "loyal" White House employees, even including, in an egregious breach of trust, Secret Service agents assigned to protect them. She will suspect (rightly) that some of her staff are confiding all they see and hear to diaries or tape recorders. Jackie, following Buckingham Palace's example, demanded a no-tell pledge from the mansion's employees—a toothless effort. Everybody from her chief of staff to the dog keeper turned out personal tales—often embroidered—from within the fortress. Her concern was understandable, but the best of the backstairs books have, in fact, provided unmatched insights and unfiltered memories that fill a gap in White House history. They bring decades of personal observations; in the case of Lillian Parks, two generations, mother and daughter, served two different worlds over a span of more than fifty years, eight White House families from the William Howard Tafts to the Kennedys.

The dean of nineteenth-century Washington journalists, Ben:

Perley Poore, ruminated that the insatiable interest in White House doings is nothing new: "The mighty public has had an appetite for gossipings about prominent men and measures ever since the old Athenians crowded to hear the plays of Aristophanes." Over his sixty-year career, Poore contributed delightful "gossipings"—on First Ladies' personalities, children's pranks, White House fashions, banquets, high-level scandals—with all the gusto he applied to national issues in over a thousand fascinating pages.

A hundred years ago Edith Roosevelt, the much admired wife of Teddy Roosevelt, commented with nineteenth-century restraint, "One hates to feel that all one's life is public property." Edith, conventional wife, mother, manager of a large and happy house, gave no interviews or speeches, allowed the fewest possible photographs, and rejected lucrative offers from publishers. Yet away from the public eye she did much to protect and enlarge the personal history of the mansion. At the other end of the spectrum was her husband's niece, Eleanor Roosevelt, who embraced the role, harnessed the White House to advance her controversial agenda, kept the mansion full of friends, and never seemed concerned about privacy, harnessing the public exposure her aunt had rejected. They were close relatives, a world apart.

Each First Lady shapes the role to her own interests. She has to. The role comes with no stated official requirements, no guidelines, no limits, no legal status. It is understood that she will be her husband's hostess at state entertainments, but that traditional wifely function has several times, over the two centuries of the White House, been delegated to a daughter, a sister, or a daughter-in-law. Nowhere is it stated that a First Lady must have a "project," but her status means that she will be the leader, always out front, never simply a regular participant. Today she would be rebuked if she limited herself to Edith Roosevelt's restricted path; yet if, like Eleanor Roosevelt, she oversteps the perceived boundaries, she will also be rebuked, as Hillary Clinton learned in heading the ill-fated White

House task force on health-care reform. Most modern First Ladies have found space for worthy and noncontroversial programs between the extremes of too little and too much. Interestingly, in the safe zone of a second term, Laura Bush is taking personal leadership of grittier social issues.

The Democratic primaries of 2004 brought the role of a President's wife into sharp focus when candidate Howard Dean's wife, Judith Steinberg Dean, a practicing physician, opted out of a public role in his campaign. Defending her decision, her husband declared that he would not ask her to be "a prop on the campaign trail"—thus denigrating active political wives. The criticism was such that Judy Dean reconsidered and soon was there on the platform in a bright red suit, close beside her husband.

Howard Dean missed the point: a President's wife is both a figure in her own right and a facet of her husband's persona. Long gone are the days when a First Lady could be passive, even virtually invisible. An aspiring President and his wife must come to grips with that reality from the outset.

By law, elected officials are given less right to privacy than other citizens, but the First Lady, neither an official nor quite the average citizen, is left dangling. The fuzzy issue was taken into the courts by Hillary Clinton's adversaries, but nothing was clarified. Meanwhile, a First Lady is subject to stinging criticism and is helpless against false reporting arising from the job she may have never wanted. But if she can handle it, she can help her husband, contribute to the country, and find her niche in history.

OVER THE YEARS the eighteen undistinguished acres surrounding the early White House had been coaxed into inviting gardens and pleasant nooks—yet who could enjoy an alfresco lunch or relax with a book when scores of office windows had a view of the lawn and, in later years, long-lens cameras intruded? Betty Ford lamented, "I love to walk out on the balcony and take in that beautiful view. But there is no privacy—there are always people at the fence with binoculars, trying to catch a glimpse." Eleanor Roosevelt, who was most at home in the President's House and was used to public exposure, was an exception—on hot days she liked to have lunch in the shade of one of the big trees close to the house. Even today, despite dense shrubbery and high fences, the electronic monitors and heavy security at every gate give a feeling of being watched, of vulnerability to cameras when taking one's ease outside.

In her three months as the first White House chatelaine, Abigail Adams deplored "not the least fence, yard or other convenience without," and even after fences were added years later, the gates were usually left open. Julia Grant came along with a solution. "I was somewhat annoyed that the grounds back of the mansion were open to the public," she said in her memoirs. "Nellie and Jess had no place to play, and I no place to walk save on the streets. Whenever we entered these grounds, we were followed by a crowd of idle, curious loungers, which was anything but pleasant." Her easy solution? Ask her husband, a general used to giving orders, to have the gates closed. He obliged.

But after the Grants left the White House, the gates were opened again, and in 1889 Benjamin Harrison groused that "not a bench or a shade tree" around the White House could be enjoyed by his family. In the sweltering days of summer, it was a real deprivation to be closed in the White House, before air-conditioning helped lessen the impact of a Washington summer, yet sitting outside was impossible, with strangers wandering in and carriages freely entering the driveway, tarrying to allow the passengers to peer into the dining room where the family was trying to have a peaceful meal.

Theodore Roosevelt did more than complain—he ordered a tennis court built for his children, away from public eyes. When critics carped about the four-hundred-dollar cost, Teddy roared, "It is impossible for them to play tennis anywhere except on the White House grounds. Only a yahoo or a very base partisan politician could object!" With that definition, the criticism ended.

Public intrusion reached a new level in Grover Cleveland's second administration, when his daughter Ruth was not yet two years old. As Chief Usher William Crook recounted the events, Ruth, in her carriage, and her nurse had been out on the lawn for "not more than six minutes" when visitors "spied them, made a rush for them, and started in to pet the baby and kiss her . . . and from that day it was impossible for little Ruth to be taken outdoors without having a group of strange women swoop down upon her."

Frances Cleveland then quite sensibly ordered the gates closed. When the public could no longer "pat the babe's cheeks, pinch its little ears, cover it with kisses, and generally maul it around," Crook wrote, the public "jumped to the conclusion that there must be some terrible, mysterious reason." As a result of trying to protect her baby from being treated like an object, rumors spread that tiny Ruth was a deaf-mute and her ears were malformed.

In the late nineteenth century, great advances in communications—telegraph, telephones, improved cameras (not to mention pushier reporters)—made the intrusion on a First Family's privacy ever more aggressive. When the press stalked Cleveland and his bride on their honeymoon, the President dispatched a stern letter to the *New York Evening Post*. The reporters, he protested, "have used the enormous power of the modern newspaper to perpetuate and disseminate a colossal impertinence," and the unseemly coverage had made American journalism "contemptible in the estimation of people of good breeding everywhere."

Long before his wedding, Cleveland had foreseen the problem of privacy. In 1885, a year before his marriage, he bought a large stone house set among woods, fields, and unpaved lanes in what was considered the country, less than four miles from the White House. (Today the neighborhood, long known as Cleveland Park, is prized for its in-town convenience.) It seemed an odd purchase for the middle-aged bachelor President, but Cleveland was wisely looking ahead to his new life with Frances Folsom, the secret love he intended to marry. After their marriage, the Clevelands carved out a unique life for a First Family, spending most of the spring and fall in their private home, spending the winter in the White House for the social and congressional seasons, and summering on Cape Cod.

Like any well-heeled Washington businessman, the President climbed into his carriage twice a day to drive to and from his White House office. When he was returned for an unprecedented second term after four years in the political wilderness, he rented a large estate nearby—this time accompanied by his pregnant wife and their baby daughter. For Frances, by then a settled matron, their "suburban" home was a lifesaver.

Almost eighty years later the Kennedys arrived, the only other White House family with an infant. Jackie followed the Clevelands' example to the extent that she could, renting an estate in Middleburg, the epicenter of the Virginia hunt country, as her retreat. It was hardly ideal for JFK, whose bad back ruled out horseback riding and who didn't care for horsey society, but Jackie, a superb horsewoman, loved the hunt. In the summer of 1963 they built a place of their own in the area, but there would be so little time left to enjoy it.

EVERY WHITE HOUSE family is caught between the opposing needs of privacy and security, sometimes even within the mansion itself. President Hoover looked up from his dinner one evening to see a stranger, properly turned out in a tuxedo, stroll through the entrance hall and into the East Room. The President summoned the Secret Service. They raced into action and quickly arrested the intruder, a midwestern lawyer who offered a simple explanation: on his way to a formal dinner, he was eager to see the White House. The city policeman at the gate, thinking the tuxedo signified a guest, welcomed him; the security agent at the front door, observing the friendly greeting, courteously nodded him into the foyer. Thus pleasantly welcomed, the stranger was examining the East Room when the President spotted him. Luckily for the man who didn't come to dinner, his briefcase contained proof of his identity and he was released.

And then there was the cheeky young couple who took a bet, in 1939, that they could crash the White House. Smartly dressed, they smiled their way in, confidently walked upstairs to the family quarters, shook hands with FDR, and made themselves at home—until Eleanor came in. Realizing they were strangers, she firmly showed them the door, and a red-faced Secret Service forthwith tightened security. That was it—no charges, no repercussions. Contrast that easy entry and easy exit with security measures demanded in the present anxious times. Guests invited to social events are first checked and show identification at the gate leading to the East Entrance, where they are rechecked against the prescreened guest list, pass through the metal detector, and ladies' evening bags may be examined.

More troubling was the California man, newly released from a mental hospital in the Eisenhower years, who climbed over the White House fence with his own ladder and, carrying a bucket of red paint, was heading for the Oval Office before he was stopped. He said he had planned to paint "I Quit" on the wall, which seemed harmless enough, if eccentric. Who dares guess at intention when the Oval Office is involved? In 1841 an attempt on President William Henry Harrison was made within the mansion itself in his one-month presidency. A lunatic entered the Red Room through a large window opening onto the South Portico. As two doorkeepers seized him, the elderly President quickly cut a window cord and they tied him up.

Who can spot an assassin? On two unrelated occasions in California, two different women who looked harmless by any standard fired close-range shots at President Ford—who was saved only by their bad aim. Then, on March 30, 1981, outside a Washington hotel a bland-looking fellow mingling with the press following President Reagan attracted no notice—until he pulled a gun and fired a near-fatal shot at the President. (John Hinckley has been in a facility for the insane in Washington ever since. He regularly petitions to be allowed to leave the grounds, a privilege strongly opposed by the Reagan children.) These attacks against particularly noncorrosive Presidents occurred after White House security had been tightened.

By contrast, the early Presidents were able to live quite normal lives, if sharing your White House family quarters with executive offices and a motley scrum of job seekers could ever be "normal." Those Presidents could move about freely: unaccompanied, they rode out on their horses along the river or into the nearby country-side and strolled the city, returning friendly greetings. Andrew Johnson picnicked with his family in Rock Creek Park, and even decades later Woodrow Wilson frequently attended the theater without causing a stir. Even in Civil War Washington, when Union troops were camped in what is now Tenleytown, a short distance north of Georgetown, and Confederate troops had thrust within twenty miles of the White House, President Lincoln regularly took Mary for carriage rides.

In today's world the most improbable freedom would be John Quincy Adams's practice of swimming, free and unguarded, in the Potomac for an hour or so on sultry summer mornings. On one exceptionally steamy day, as journalist Perley Poore reported at the time (contrary to some later apocryphal inventions), the President had his steward row him and his son John across to the Virginia side for better swimming. In midriver they slipped out of their clothes, eager for the plunge, when a fierce gust swamped their little boat. The three men "were forced to abandon it and swim for their lives to the Virginia shore," Poore reported. The steward, salvaging their few garments, set out for help. For three hours the very proper and very bare President, trying to keep out of sight, alternately paced the shore or stayed well below the river surface, until at last the steward returned with a carriage and clothing. Later, the President recounted his tale of being caught wrapped in nothing but his thoughts, to which his British guest replied, "I now have a clearer idea of republican simplicity!" A modern President could not dream of such freedom, and today's Virginia commuters, fighting the rush hour on the Potomac bridges, can only smile at the mental picture.

Almost as incredible to present-day Washingtonians are accounts of President Taft and his personal aide, the very proper Major

Archibald Butt, swimming in Rock Creek, which flows through Washington's busiest in-town parkway. Doubtless, they were correctly attired for the occasion, if there is "correct" attire for a President, in this case a man too big for a bathtub, splashing in the midcity stream.

ANNOYING AS IT can sometimes be, White House families understand the inherent danger of being who they are; they understand the constant presence of the Secret Service beyond their protected eighteen acres. And yet all of them, at one time or another, wish that they could sometimes go about as themselves. When Michael Reagan asked his father what he found most difficult in the White House, the President spoke for generations of his predecessors: "The hardest part is that I wish I could walk right out the door . . . and just walk down Wilshire Boulevard like I used to, and have nobody approach me, no Secret Service, and just window shop. That has been taken away from me for the rest of my life."

Reagan might have envied Cal Coolidge, who did just that in downtown Washington. He tarried at the best shop windows on F Street, now and then bought Martha Washington chocolates for Grace, and seemed not to have been noticed at all. Window-shopping, Coolidge said in one of his rare comments, "takes me away from my work and rests my mind." Not that he is ranked as one of our more productive Presidents. His Secret Service agent, Colonel Edmund Starling, remembered the President "enjoying himself like a small town boy strolling down Main Street on Saturday night." The difference between the two chief executives strolling along a city street was that Reagan's expansive personality filled the air around him—he would have been noticed had he been an off-stage prompter—while Coolidge's cramped spirit made him seem less than the man in the White House.

Each new presidential family comes into the White House hoping to remain "normal." Eleanor Roosevelt declared, "I'm just going to be plain, ordinary Mrs. Roosevelt"—a contradiction in terms. "I'm not going to be trailed around that way. Nobody is going to hurt me. I'm not important enough." Though no First Lady has ever been physically attacked, Eleanor, a social activist decades ahead of her time, was viciously abused verbally and undoubtedly received many warnings. Over the years more threats have been made against a President's wife than his Vice President, and more than once the Secret Service has required a First Lady going into a volatile situation to wear a bulletproof vest.

First Families' efforts to step unaccompanied outside the White House cocoon, on their own, meet with little success. Bess Truman planned to continue driving herself around town as she had for ten years, but she caused such a stir—other drivers honked and waved, pedestrians approached her car for an autograph when she was stopped at a red light—that she gave in, insisting that she be driven in the least conspicuous White House cars. Barbara Bush also used the smallest cars, and when she asked to travel by commercial plane the Secret Service nixed her request as too dangerous.

When a First Lady ventures out of the White House in search of normality with personal friends, she knows she will read about it in the chatter columns the next day. (Laura Bush lunches in Georgetown. *The Washington Post* reports: "very friendly and looking much younger than on TV, ordered gazpacho and a chicken sandwich, her friend picked up the tab and tipped generously.") Nancy Reagan occasionally went to lunch with political columnist George Will, a Reagan favorite, in a quaint village off the beaten track in Virginia, where the country restaurant still proudly displays the record of her visits, along with a photograph. Did this ultimate White House source share White House secrets with the columnist? Maybe, maybe not, but a little insiders' political gossip was likely on their menu. First Ladies need a break from being First Ladies.

———

PITY A PRESIDENT'S teenage daughters, under the watchful eye of the Secret Service every minute they are outside their home. Luci Johnson, a White House teenager who grew up to be the mother of four teenagers, now laughs, "It's every teenager's mother's dream— and every teenagers' nightmare." Her sister, Lynda Robb, interjects, "It's like having your big brother on a date with you," and adds, rolling her eyes, "The worst thing is having your 'brothers' go along on your honeymoon!"

Though White House children grumble about their ever-present "nannies," they have come to understand the necessity, especially in the current security climate. For several days in 1974 Susan Ford, daughter of the then Vice President and a virtually unknown public figure, was ordered not to set foot outside her home. Despite her complaints, the order was immutable. She later learned that her name was third on a list of targets drawn up by the Symbionese Liberation Army, a domestic terrorist group. The name ahead of Susan's was Patty Hearst, the daughter of the powerful Hearst publishing family, who was kidnapped by the SLA in California, held for months, brainwashed, and involved in a notorious murder committed during a bank robbery. Susan now tells of the Secret Service's adamant orders with gratitude and a shiver.

A generation after those daughters, the problem is the same—or worse—for the Bush twins, who have reaped their share of headlines. After Jenna's first arrest, Laura Bush protested the news coverage, insisting that it was "a family matter," an understandable but unrealistic reaction, since the case was heard in open court. "We wanted them to have as normal a life as possible," Laura explained to the *Washington Times*. "To get in trouble, if they'd get in trouble. Obviously we didn't want them to get into trouble, but we wanted them to not always feel the pressure from their peers of having their parents in the public eye."

In the Texas governor's mansion the twins had come through six years of middle school and high school virtually unscathed, but they knew that in Washington, with its enormous press corps, they would come under a much more powerful magnifying glass, just as they were entering college and wanting a more independent life. Camp David was the answer when they absolutely had to go to Washington; the Crawford ranch offered a relatively normal home life with their parents, though not much of the lively young scene they like. Traveling in Europe with friends has allowed them to go wherever they liked virtually unnoticed, easily blending in with other trendy young vacationers from all over the world, none of them looking for the American President's daughters, who look not at all like twins.

Even Jackie Kennedy, whose face could trigger photographers' shutters around the world, discovered that she could slip out of the White House anonymously. Staying below the radar of publicity, she would don an old raincoat, jeans, and a sweater, tuck her hair under a scarf, and take Caroline and John to a puppet show or the circus. At the last moment they would slide into their seats for a fun afternoon, no different from the kids all around them. She loved doing that with her children—but she surely took extra delight in having evaded the vultures of the media.

Shielding her children from the media became an obsession with Jackie. From the moment the Kennedys appeared on the scene, America was smitten. Caroline, a sunshiny blonde, part angel, part imp, was only three; baby brother John, not yet three months old, was not much more than a bundle. With one exception, they were the youngest presidential children to live in the mansion—the Grover Clevelands' daughter Ruth was eighteen months old when her father returned to the White House for his nonconsecutive second term, and baby Esther was born in the White House, the first for a President.

Jackie's determination to keep photographers out of range of her children was an irony in that Jackie Bouvier, as the inquiring photographer for the *Washington Times-Herald,* more than once ambushed the children of officials—including then–Vice President Nixon's six-year-old daughter Tricia—to get a story. In the White House, however, Jackie deplored the media—except when she wanted coverage—and resented the pushy public. Protesting "I'm sick and tired of starring in everybody's home movies," she wanted the South Lawn screened by rhododendrons, a plan overruled by her husband, who insisted that tourists should be able to see the President's House, and by the White House police, who called the plan a problem. "If Mrs. Kennedy had her way," Chief Usher West commented in retrospect, "the White House would be surrounded by high brick walls and a moat with crocodiles."

The President conspired with the official White House photographer to circumvent Jackie's iron will: when she was out of the country he brought photographer and children together in the Oval Office. Those stolen moments captured images—John peeking out from under his father's desk, Caroline dancing with him and shuffling in her mother's size-eight pumps—that became cherished memories of the fleeting "Camelot."

IN DECEMBER 1963 Lynda Bird and Luci Baines Johnson, the first teenage daughters in the White House since the blithely unmanageable Alice Roosevelt fifty years before, instantly became public property. Whatever they did—the way they dressed, their hairstyles, their makeup (all of which are crucial matters for teenage girls)—became fair game for comment and criticism. "The life of a family of goldfish is a secret compared to someone growing up in the White House," Lynda once said ruefully. "The President's daughters must be an example to other young Americans." But both daughters withstood the scrutiny without mishap. Years later Lynda observed, "Mother used to tell us, 'Don't do anything you wouldn't mind seeing on the front page of the newspaper.' I used to complain, 'It's not fair,' but that's the way it is. We recognized that you have to watch out, for your parents' sake—and for your own sake."

Not long after moving into the White House, sixteen-year-old Luci mused, "You can be miserable or you can be happy. Sometimes I hate it. It's a museum. Eighteen thousand people come through my house on a holiday." Lynda, who was dating, breaking up, dating more, chimed in, "Wherever we go, whatever we do, we may end up in the newspapers. I never have any privacy. I don't necessarily think I owe my life to the American people." A freshman at the University of Texas, Lynda had found it impossible to be just another student, with the media dogging her and the Secret Service posted at her dorm—and her mother was gently reminding her history-loving daughter that living in the White House "was an experience I shouldn't forgo."

"With great reluctance" she rejoined the family at the White House, bringing along a Texas friend, Warrie Lynn Smith, as her roommate. Soon Lynda's friendship with handsome young Holly-wood bachelor George Hamilton became an item in gossip columns coast-to-coast. Lynda's visits to her newly discovered world of young Hollywood stars marked her emergence as the striking beauty she remains now she is past sixty.

Saucy Luci found a way to thwart the media: for a fun weekend at Marquette University (where she met the fellow she would marry) she tucked her raven hair under a blond wig and hid behind a fake name, "Amy Nunn." To this day she delights in telling how she fooled the press: "No one noticed 'Amy'—they were looking for the President's daughter." If there is an Amy Nunn out there some-where, know that you played a role in a White House courtship.

Lynda recently revealed her own adventure in disguise: "I was cu-rious about what White House tourists were saying, so I put on a trench coat and a scarf over my head and joined a regular tour. (They weren't saying much.) And when they left, I just walked out the door with them, and onto Pennsylvania Avenue. Not a Secret Service agent in sight! But then I thought better of what I was doing and talked the officer at the gate into letting me back in." Her little escape from the White House was like a runaway puppy that scratches to come back in.

THOUGH WHITE HOUSE daughters, almost without exception, recoil from media attention, there was one who reveled in it. A century before the recalcitrant Bush twins became presidential daughters, there was Alice Roosevelt, who relished shaking up the Washington scene and, unlike Jenna and Barbara, basked in the attention she attracted. From her first day in the White House, Theodore Roosevelt's beautiful, outrageous daughter was the talk of Washington.

Born into the American aristocracy, "Princess Alice" was the center of whatever she chose to do, the star of every party. A popular song was written about her; a color, "Alice blue," was named in her honor. She danced until all hours, smoked in public, flirted blatantly, placed bets at the races, dived in a submarine, raced a car from Newport to Boston at a hair-raising twenty-five miles per hour, and was seen in an automobile with a gentleman—without a chaperone! Fashionable ladies and shopgirls alike devoured every morsel of her social whirl, what she wore, how her hair was done. Her close friend and later rival, the young Countess Marguerite Cassini, put it simply: "The country fell in love with her."

When a friend of the President, concerned about Alice's latest antics, advised TR that he really must do something to control his daughter, Roosevelt, by his own account, replied: "I can either run the country or attend to Alice, but I cannot possibly do both." Being President was easier. He could face down wild beasts or tame a hostile Congress, but when it came to Alice the daring Roosevelt admitted defeat.

During a summer vacation at Sagamore Hill, a young man drove up to the gate demanding to be admitted—he was coming to marry

Alice. The Secret Service wrested his pistol from him and reassured the President that the intruder was not an assassin but a lunatic. TR accepted the explanation calmly: "Of course he's insane—he wants to marry Alice."

When the family was thrown into the White House overnight, Alice admitted in her memoirs, "I was not particularly elated by the fact that Father had become President. It hardly even occurred to me that it might be interesting." At seventeen, she was much too engrossed with her fashionable young crowd and, in particular, herself: "I was not merely an egotist, I was a solipsist," she blithely confessed. "No young person could ever be more frivolous and inane, more scattered and self-centered than I was." No one felt inclined to disagree.

She soon discovered that the White House was the center of the social universe: on January 3, 1902, the mansion was the splendid setting for her debut, a ball attended by seven hundred guests. When her parents refused to give the lavish favors customary in her set— gold watches, silver cigarette cases, and the like—and ordered fruit punch instead of champagne, the imperious princess pouted at the "horrid blow to my pride." But when the Marine Band expanded its repertoire to include her favorites—"The Debutante Waltz," the "Hop Along Sing" two-step, and "Dem Goo Goo Eyes" polka—she recovered. She loved the front-page headline: HISTORIC MANSION SENDS FORTH ITS FIRST DEBUTANTE.

After she was properly launched in society, Alice began seeing a great deal of bachelor congressman Nicholas Longworth, fifteen years her senior, a superb dancer, talented violinist, and bon vivant whose motto was "I'd rather be tight than President." And, not a minor point, he was the scion of a wealthy blueblood Cincinnati family, which made him a suitable match for the daughter of the President—*and* a Roosevelt. Nick was also quite the ladies' man (as his later life would attest), seriously courting both Alice and her best friend, the glossy young Countess Cassini. The competition between the princess and the countess undoubtedly stoked Alice's determination to win.

The President may have intentionally fostered the romance by including Alice in the large official party making a four-month goodwill tour of the Far East—and Nick Longworth was in the group. The newspapers were titillated: "Alice in Wonderland," one blared. "Representative Longworth to Go Along—Tropical Romance Anticipated."

Once on board ship Alice let everybody know she was there by jumping into the swimming pool—fully clothed, of course. While most of her traveling companions were shocked, Nick found the untamed Alice delightful, and by the time the leisurely mission was over, they were engaged.

Her White House wedding in the East Room was the social event of the era; the 680 guests, the cream of East Coast society, arrived for the high-noon ceremony in 145 carriages, requiring extra police and a platoon of door openers. The huge room was a sea of Easter lilies in mid-February; Alice was regal in ivory satin with an eighteen-foot-long train of silver brocade. Unlike most White House brides, she had no attendants—Alice was not one to share her spotlight.

Her wedding gifts were fit for a bona fide princess—the Empress of China sent eight rolls of high-carat gold brocade, a white fox coat, an ermine coat, two rings, earrings, and rare white jade; the King of Italy, a marble-inlaid table; the French government, a Gobelin tapestry. The government of Cuba chose a string of precious pearls that became her signature. Gifts poured in by the thousands from plain Americans, including such utilitarian objects as a mousetrap, a feather duster, and a box of snakes.

The marriage lasted, of course, divorce being unthinkable, but with what degree of happiness? Longworth, Speaker of the House of Representatives and one of Washington's most powerful men, was depicted in print by the House doorkeeper as being "one of the greatest womanizers in history on Capitol Hill." He was also a heavy drinker.

The rumor mill linked Alice with the dynamic Senator William

Borah of Idaho, the speculation extending to her daughter, Paulina. Washington society whispered that the night before Paulina's wedding, Alice told the young bride that her doting father was in fact not her father—and offered no further explanation. Her coldness toward her daughter was well-known, but what purpose was served by delivering such a shock at that moment? When Paulina was thirty-one, she fatally overdosed on alcohol and barbiturates; Alice then brought up her granddaughter, Joanna Sturm, Paulina's only child, with affection she had never before revealed.

For eight decades Princess Alice brightened the capital. A coveted invitation to her faded Embassy Row town house, a few blocks from the White House, was a passport to her unique world that linked long-ago times and current players. No one could touch her record—beginning when she was a small child, she had met every president from Benjamin Harrison to Jimmy Carter and "had a memory" of Mrs. Grover Cleveland, "young, lovely, friendly," in the White House.

The most famous article in her cluttered drawing room was her signature pillow, exquisitely needlepointed: "If you haven't got anything good to say about anybody, come sit next to me." Always the center of attention, she lightly tossed off politically devastating, cyanide-tipped barbs: "The Hoover vacuum cleaner is more exciting than the President—but, of course, it's electric"; Republican candidate Wendell Willkie, the Wall Streeter running as a plain fellow from Indiana, "sprang from the grass roots of the country clubs of America." Her select audience roared at her perfect imitation of her first cousin, Eleanor, whom she found a tiresome do-gooder, and she dismissed her more distant cousin, Franklin, as "one part Eleanor and three parts mush." Eventually her personal attacks became too much for the cousins, who banned Princess Alice from the White House she had once dominated. But after the FDR days she was back again, in full sail, for big events and all of the weddings.

After her death in 1980, at ninety-six, Washington seemed a little duller; in all of the dozens of White House families there has never been a figure to equal Alice. The distinguished columnist Joseph Alsop summed her up: Alice Roosevelt Longworth was "Washington's other monument."

THIRTY-TWO YEARS before Alice's spectacular wedding, the nation had been beguiled by the international marriage of President Grant's adored only daughter, eighteen-year-old headstrong Nellie. On her return voyage following a social whirl in London, Algernon Charles Frederick Sartoris, a handsome and wealthy Englishman, had swept her off her feet in a shipboard romance.

Despite her father's opposition to the match, Nellie would, as always, have her way. In an East Room extravaganza on May 21, 1874, Nellie, surrounded by eight bridesmaids, shone in a gown of white satin and lace that cost five thousand dollars. The dazzling wedding gifts included a diamond necklace and earrings, along with a check for ten thousand dollars from her parents, and an ode to the bride composed by Walt Whitman, America's foremost poet and a friend of the President. Its closing line: "O, bonnie bride! Yield thy red cheeks today unto a Nation's loving kiss." The wedding breakfast menu, featuring twenty-five gourmet dishes, was inscribed on white satin for the two hundred guests.

There was one unique note: the bridegroom carried his own bouquet of orange blossoms with a tiny silver banner embellished "Love." Small wonder that the tough old general's eyes filled with tears. An aide later found him sobbing in his daughter's room, his head buried in her pillow. Father knew best: after a few years and three children, the unhappy Nellie left Algernon and came home to America and Papa. (She later married an assistant postmaster general in a quiet ceremony in Canada.) Nellie has plenty of company: there is a high rate of failure among the marriages of presidential daughters—the glamour of the White House may be blinding.

Margaret Truman, the first teen-age daughter in the White House since Helen Taft, was clear-eyed about that glamour: she was determined not to become engaged while her father was President. How could she be sure whether a suitor was entranced with herself or the White House? "I can't wait to get out of here!" she told her friends. Shortly after graduating from college in 1947, she moved to New York for a life on her own, commuting to Washington for weekends and big events. In 1956, four years after the White House was out of her life, she married Clifton Daniel, a *New York Times* editor. The ceremony took place far away from the glare in the little Episcopal church in Independence where her parents had pledged a lifetime of devotion thirty-seven years earlier. This White House daughter's marriage was long and happy with the husband who looked like Leslie Howard and their four sons.

But what bride thinks that she won't live happily ever after as she chooses her gown and frets over the decorations topping the cake? For any lovestruck maiden, the wedding must be her vision of perfect, but a President's daughter has more to consider: hers will be an event that transcends family scrapbooks to become part of White House history. They are an elite group—in more than two hundred years and among the one hundred or so presidential children who lived in the White House, only twelve daughters have married while their fathers were in office, eight in the White House, the other four in four quite different settings. And not to be overlooked is the one son who wed there—John Quincy Adams's son John.

In her father's first year in the White House, 1913, Woodrow Wilson's middle daughter, Jessie, had thought to have a simple wedding in the smaller Blue Room, until the President's burgeoning guest list filled the East Room. The nuptials almost went on without the bridegroom—Francis B. Sayre, a young academic, arrived at the White House gate in full dress, ready to claim his bride, but without an invitation in hand. The guard at the front gate refused to admit him—only when the police captain was summoned was the groom cleared to attend his own wedding. (He was a serious young man who later became dean of Washington Cathedral.)

If Jessie had thought to keep the event low-key, she failed; the following day the wedding blanketed virtually the entire front page of *The Washington Post*. A three-line banner headline declared, NATIONS OF ALL THE WORLD DO HOMAGE TO THE WHITE HOUSE BRIDE AS SHE TAKES SOLEMN VOWS AMID SCENE OF UNEQUALLED SPLENDOR, and a photograph of "The Nation's Daughter-Bride" spanned a quarter of the page. Sidebars covered such news as "Bride Cuts the Cake; Sister Gets Ring; All Dance the Tango."

It was suitable—probably fixed—that her sister Eleanor should get the slice with the ring. At twenty-three, "Nell," the youngest of the three Wilson daughters, was unofficially engaged to a nice young man before her family moved into the White House. But in that heady new atmosphere the young fiancé was soon past tense, and less than six months later, President Wilson was giving another daughter in marriage. This time great titillation centered around the bridegroom—Secretary of the Treasury William Gibbs McAdoo, a much older widower of great charm and great wealth, for whom many Washington ladies' caps had been set. And this time wedding coverage filled the entire front page of the *Post*. It was a brilliant beginning that ended in divorce twenty years later.

In August 1966 Luci Johnson, a Catholic convert, bypassed nearby St. Matthew's Cathedral, where she had been baptized, to choose the largest Catholic church in America for her wedding, the National Shrine of the Immaculate Conception, in Washington, with the archbishop officiating. It was the first church wedding for a presidential daughter and the most elaborate of all White House weddings—twelve attendants walked down the long aisle, a men's choir of one hundred fifty voices sang, and a huge reception followed in the East Room. And, of course, there was the very young bridegroom, Patrick Nugent. A week before the event, Luci demonstrated that although she was barely nineteen, she was in command: she held an open press conference, answering all questions and sharing personal insights with astonishing aplomb. But that thoughtful beginning augured no better success for the marriage than Nellie's impulsive decision. After Patrick completed his tour of duty in the Vietnam War, the Nugents moved to Austin, Texas, had three daughters and one son— who has always been known by his distinctive middle name, Lyndon. And then the story ended in divorce; much later, Luci's marriage to Ian Turpin, a British banker, took her away from Texas to Toronto for several years, then back to Austin where they are civic leaders.

Lynda Johnson's wedding in December 1967 was the ultimate White House story from first hello to romantic "I do." Charles Robb, a handsome Marine officer, was a military aide at the White House, where familiarity breeds romance. Since Chuck was regularly assigned to the mansion, the couple could court in the solarium with the sleuths of the society press none the wiser, until rumors ballooned into fact. They were wed in a Christmastide ceremony in the East Room, a marriage that carried the impact of the White House through Robb's term as governor of Virginia and his twelve years in the Senate.

The two Nixon daughters who followed the Johnsons chose surprisingly different paths. In 1968, a few weeks before her father's inauguration, Julie Nixon married David Eisenhower in New York, obviously to avoid the hordes of press and endless demands that come with a White House wedding—especially one that combined two such famous names.

The surprise lay in the personalities of the two sisters. Julie, a warm and lively brunette, had always been outgoing; Tricia, the elder daughter, a lovely blonde, was perhaps the most reclusive of any modern-day White House daughter. Even within the family quarters of the mansion, she spent much of her time in her suite. Yet it was Tricia who added a historic page to the White House scrapbook—she married Edward Cox, a young New York lawyer, in the only wedding ever held in the Rose Garden, complete with a tent for the pack of reporters and full television coverage.

It had rained off and on throughout the morning, and minutes before the wedding march was to begin the skies still glowered. The family, the White House staff, and the nation's top meteorologists agonized: move the wedding into the East Room or take the chance? Tricia made the bold decision: take the chance. Presto, the sun broke through, blessing a perfect June day for vows in the flower-entwined gazebo.

The George H. W. Bushes' daughter Dorothy was a very contemporary young woman—a working mother of four children, and divorced, who lived in her own home and managed to stay out of the photographers' range. Seeking a quiet setting for her second marriage, to Robert Koch (a Democrat!), "Doro" found just the place— Camp David, the elegantly rustic presidential retreat in the Maryland mountains. Hers is the only wedding held at that perfect hideaway. An earlier daughter determined to have a quiet second wedding, FDR's Anna had a private ceremony in the family's Hyde Park home. With careful planning, the press can—mostly—be escaped.

Since the Bush twins are in their mid-twenties, it is not unreasonable to presume that one or both will be married during their father's second term. With their flair for independence and their savage dislike of the press, their choices of venues and style—not to mention bridegrooms—will be followed in every detail by the media they detest.

ALTHOUGH FROM THE earliest days journalists and White House families have been locked in an uneasy relationship, sometimes serving mutual advantage, more often as adversaries, there was a time when the relationship was easier. Imagine today's photographers agreeing not to show Franklin Roosevelt's leg braces or reporters never mentioning his struggle to propel himself—even when the brace locks gave way and he fell at a public appearance. Imagine Grover Cleveland, aboard a friend's yacht in 1893, having two secret operations for cancer of the mouth, requiring an artificial jaw of vulcanized rubber, without as much as a published rumor. Imagine Edith Wilson's "presidency."

Contrast that secrecy with the practice of almost anatomy-course briefings on the President's health, beginning with Eisenhower's heart attack, ileitis, and stroke, escalating with Lyndon Johnson proudly displaying his incision after gallbladder surgery and descriptions of every mole and polyp since. Yet, it should be noted, John Kennedy's Addison's disease, a serious and chronic medical problem, was not addressed by his doctors, despite the fact that it was widely known. In the Kennedy White House the press often observed the President's puffy face, which they attributed, among themselves, to cortisone treatments. However, Kennedy never seemed to be impaired, and the disease itself was rarely mentioned.

In 1974 the forthright and fearless Betty Ford made the decision to be completely candid about her breast cancer. In an act of personal generosity, she allowed details of this deeply personal operation to be made public, demystifying the treatment and setting an example that would erase the lingering embarrassment of cancer. In an interview on *60 Minutes*, the First Lady explained her thinking: "If I had it, many other women had it. If I don't make it public, then their lives will be gone."

She was still on the operating table when the White House set up a press briefing with cancer specialists who dealt with every detail; no question went unaddressed. She made her experience with cancer a teaching tool for all other women, and her openness made it easier for Nancy Reagan when she had a similar diagnosis. The thinking in the White House and the candor of the First Lady had come a long way since Cleveland's clandestine "tooth extraction."

———

ELEANOR ROOSEVELT, WHOSE column, "My Day," ran in many newspapers, was the first to see the press as a useful tool to advance her many concerns. She instituted weekly news conferences for women journalists, who then had virtually no opportunity to cover the President's daily scramble around his desk. In 348 such meetings, Eleanor handled a free flow of questions and was never caught off base; she knew what she stood for and said so—and deftly produced news stories that she wanted to receive attention. An early, and successful, news manager.

Women reporters have not always been as agreeable as those Eleanor cultivated. One of the first and worst among Washington's earliest correspondents was Mrs. Anne Royall, an enterprising, self-declared journalist, who in about 1825 bought herself an old printing press, hired a couple of drifter printers, and published a small weekly sheet called the *Huntress*. She called on "every person of any distinction." If they agreed to see her, they received good press; if not, Perley Poore recorded, "she abused them without mercy." Scruples were not her specialty. John Quincy Adams derided her as "a virago-errant in enchanted armor, redeeming herself from [poverty] by the notoriety of her eccentricities," which apparently kept her newspaper afloat.

With obvious delight, Poore reported that "Mrs. Royall's tongue at last became so unendurable that she was indicted by the Grand Jury as a common scold." Found guilty, she would have been sentenced to the ducking stool had her lawyer not gotten her off with a fine. The ducking stool for troublesome reporters might have considerable appeal to today's White House residents.

six

All in the Family

———◆◆◆———

A NEW FIRST family moving into the White House discovers that it is more than an awesome dwelling; it is an almost living force that will affect their lives forever. There is an interlocking between the entity that is "the White House" and the family as they share such bright moments as weddings and tragic times of death. Some families loved it as their own; others were eager to leave.

As the first Roosevelt family packed up after almost eight exhilarating years, Theodore Roosevelt wrote to a son, "I do not think any two people ever got any more enjoyment from the White House than Mother and I." Julia Grant had also reminisced in superlatives: "My eight years in the White House were the happiest time of my life." After four years out of office, both TR and Grant sought another term, but by then each had lost his luster and was rejected. Grant was denied his party's nomination; Roosevelt ran on a third-party ticket and was rebuffed by the voters.

At the other extreme were those who could not adjust and were eager to escape to privacy, away from the anxiety about making a wrong step, saying the wrong thing, and worrying how the attention would affect their children. Elizabeth Monroe, Peggy Taylor, and Eliza Johnson all eased their nerves by bequeathing the First Lady's role to a daughter and retreating to the family quarters.

Elizabeth Monroe, a New York socialite who had cut a swath through diplomatic circles when her husband was minister to France, appeared only at major events; as a result, White House dinners largely became stag events. Ever-present chronicler Margaret Bayard Smith commented that the Monroes, after almost two years in the White House, "are perfect strangers not only to me but to all citizens." Peggy Taylor, the daughter of a prominent Maryland family who had spent her adult life as an army officer's wife on the rough American frontier, regularly went out to church and entertained personal friends in her own rooms but shunned all official entertaining. Eliza Johnson, who was in poor health, was know to have ventured downstairs twice—to attend a party for her grandchildren and to greet the exotic Queen Emma of the Sandwich Islands (Hawaii), the first of generations of royal visitors to call at the White House.

The congenial, gregarious George and Barbara Bush, who had rarely been in the family quarters during his eight years as Reagan's Vice President, immediately transformed the White House into a family home. Throughout the inaugural weekend of 1989 the Executive Mansion overflowed with Bushes, Walkers, and Pierces— relatives from every branch of the extensive family tree. (Another member of that tree had inhabited the White House: in 1853 Franklin Pierce served as the fourteenth President, a Democrat as amiable as the forty-third President, a lateral kinsman several generations removed, through Barbara Pierce Bush.) Throughout their term the Bushes enlivened the family quarters with children, grandchildren, and dogs. "Anybody who couldn't be happy in the White House couldn't be happy anywhere," Barbara declared.

Laura Bush agrees. "People would be surprised how real life in the White House is, and how normal," she said during her third Christmas holiday there. "I mean, it's magnificent, and we live with furniture and art that's literally museum quality, with a wonderful staff and a chef and all those things that are true luxuries. We have friends over for holiday parties—nearly every bedroom is filled with friends."

Still, in her memoirs even Barbara confessed to mixed feelings as early indications began to look grim for her husband's reelection in 1992. "I'd hate to have George lose, for his sake," she thought, "(but) think how good it would be for our children. They could get on with their lives." And indeed they did get on with their lives—with two governorships and two terms in the White House. In February 2005, shortly after their son's second inauguration, perhaps the most meaningful unofficial event in White House social annals took place: President–son honored President–father and First Mother at a dinner for more than a hundred of their friends, celebrating the sixtieth anniversary of a perfect marriage. It was a unique blend of personal and historic in what, that night, was the Presidents' House.

Despite its unequaled amenities, the White House could not mend the dysfunctional Reagan family. Patti Davis termed her step mother harsh and her father "indifferent." Son Ron wryly commented, "Our family was famously fractious, and it stayed fractious," but he never spoke unkindly of them in public and was the strong arm for his mother during his father's state funeral ceremonies.

Martha Washington once commented philosophically, "I have learned from experience that the great part of our happiness or misery depends on our dispositions and not on our circumstances," an observation that still rings true. Martha's insight was never truer—as to the misery—than with the John Quincy Adamses after they attained the White House. In her journal Louisa Adams lamented, "There is something in the great unsocial house which depresses my spirit beyond expression and makes it impossible for me to feel at home or to fancy that I have a home anywhere." That she quickly recovered following her husband's defeat for reelection in 1828 suggests that the White House itself was at least partly to blame for her problems.

Surely her dyspeptic moods were worsened by the problems within the family. In the long history of the White House only one son has been married there, and that marriage left the family in shreds. Louisa had brought into the family an orphaned niece, Mary Catherine Hellen—a little girl who grew up to be a flirtatious young woman. She first captured the heart of the youngest of the three Adams sons, Charles, then shifted her wiles to the eldest, George, and ultimately dumped him for the middle son, John Adams II. The parents saw the whole episode as a disgrace, but the couple went ahead with their wedding in the Blue Room. Neither John's mother nor his two brothers attended; the President, papering over the embarrassing family split, attended and a groomsman reported, "unbent for the nonce and danced in a Virginia reel with great spirit at the wedding ball." Did the little vixen make the right choice of the brothers? George became an alcoholic wastrel; John, who won the maiden, died at thirty-one; Charles, the brother she only toyed with, lived to a ripe and respected old age.

Somewhere between the extremes were the Trumans, who found Blair House, where they lived while the White House was being rebuilt, more to their liking than the Executive Mansion. It was much smaller and entailed fewer official demands. The way they lived, Chief Usher West observed to Truman biographer David McCullough, the mansion "might as well have been in Independence. As far as everyday living goes, they were no different."

IMAGINE THE TREPIDATION of Frances Cleveland, at twenty-one the youngest of all First Ladies and one of the prettiest, arriving at the White House as a bride who had to become a grown-up wife and proper hostess overnight. Chief Usher Ike Hoover told of a touching little scene he witnessed: the new First Lady had come upon White House servants dancing in the library and discovered the fireman, a German of considerable talent, "banging away on the piano"—all quite against the rules. Embarrassed and nervous, they stopped when the First Lady walked in.

"Did she rave and discharge [them]?" Hoover asked rhetorically. "Not at all." A bit wistfully she said, "Please don't stop," and asked if she might stay and watch. It had been only a short time before that she, too, had been a carefree college girl; now she was married to a President more than twice her age and thrust into White House life with all its constrictions. "Frank"—the First Lady—sat and looked on, surely feeling a twinge of envy at the young servants' rollicking fun.

Frances was a reference point, possibly a role model, for Jackie Kennedy. Both were mothers of small children, both escaped for long periods in their private homes, both organized a little kindergarten in the White House, both were disturbed by intrusions on their children's lives—and each gave birth to a child while she was First Lady.

Long-serving Chief Usher Hoover, a stern judge of First Ladies, immediately fell under Frank's spell, as J. B. West did under Jackie's. Ike Hoover found his First Lady "brilliant and affable. . . . Her very presence threw an air of beauty on the entire surroundings." Her entertaining introduced a new generation's thinking—some receptions were held on Saturday afternoon so working women could attend. Who had ever thought of them before?

By the time the Clevelands returned to the White House for a second term, the young bride—more self-assured, the slender waist a bit more rounded—had become a mother and then gave birth to a second daughter in the White House. For encouragement in making the White House a home, Frances might have looked back to Jefferson, a widower in a still-raw mansion who created a happy, fun-filled atmosphere lively with children. (Four of his own six had died in early childhood.) Since both of his surviving daughters were married to congressmen in their father's early years as President, they could bring their children—ultimately eight—and live much of the time in the White House. Daughter Martha Randolph won an asterisk in White House history by giving birth there to its first child, a boy christened James Madison for her father's great friend and successor.

Jefferson doted on his grandchildren, loved the sound of their laughter, read to them, lavished them with gifts. Years later granddaughter Ellen Randolph reminisced, "He could see into our hearts. He gave me my first silk dress, my first good writing desk . . . all the joyful surprises of my girlish years"—a watch, a saddle and bridle for her horses, her Bible, her Shakespeare.

Creating a hospitable White House, Jefferson entertained at dinner two or three evenings a week—sharing good food, lively talk, and French wine that cost him dearly. Alone, he relaxed with his books (his personal library became the genesis of the Library of Congress) and played his violin. Bracketed by two strong, much admired First Ladies, the first single President equaled them in making the White House a home.

Twenty-eight years later, grieving widower Andrew Jackson also brought an extended family with him, creating a White House full of children and fun. His beloved Rachel's nephew, Andrew Jackson Donelson, was his private secretary, and Donelson's twenty-one-year-old wife, Emily, became the President's hostess; they came with one young son and added three babies. With two Donelson nieces and a friend's daughter, Old Hickory was surrounded by beguiling babies and young beauties, who helped lift the shadow of Rachel's death. He also welcomed the bride of his adopted son, Andrew, with a grand reception and a series of dinners. (They returned to Tennessee to manage Jackson's plantation—poorly, it turned out.) In a new role, the old soldier played Cupid to the young ladies and gave two of them White House weddings; he arranged four christenings, and for the little girl named after Rachel, he ordered, "Spare no expense nor pains. Let us make it an event to be remembered." His gift for baby Mary Rachel: her own eight-year-old slave.

Again defying stereotype, the tough old general transformed the rather stiff, threadbare John Quincy Adams White House into what historian Amy La Follette Jensen termed "something both beautiful and awe-inspiring," elegance imported from Brussels and Paris. The traces of the patrician New England Adams dynasty were swept

away by a man of no antecedents from the nation's barely tamed
frontier.

WHO WOULD HAVE guessed, hearing Julia Grant's superlatives
about life in the White House, that initially she did not want to live
there at all? She very much wanted to be First Lady but preferred to
continue to live in the Grants' own home, a mansion that had been
bought for the general by admirers and was located just a few blocks
from the President's House. But when her proposal was turned
down, she bowed to tradition—and made the Grant White House
the sunniest of homes.

For a number of years Julia had moved from one army post to the
next to be with her husband, taking along which children she could,
leaving the others with relatives. For Grant, who had been a heavy
drinker when his wife was not around, life in the White House must
have met a fundamental need, for there were no more whispers
about his penchant for "John Barleycorn." After their haphazard
army life, there was joy in such simple acts as sitting down to meals
together. Chief Usher William Crook (an unfortunate name, given
the scandal-ridden Grant administration) recalled their "jolly family
dinners" when the President and the children, roaring with laughter,
threw bread bullets at one another.

In that loving and lively family home, a gentler side of the war-
toughened President emerged. "He never told an indelicate story,"
Perley Poore reported, "and his strongest language was 'dog on it!'"
His features "softened"; gray crept into the brown hair and beard, and
thirty-eight additional pounds were a tribute to White House life.
(Crook listed the Grants' typical "plain" breakfast: "broiled Spanish
mackerel, steak, bacon and fried apples, rolls and buckwheat cakes.")

Trying to support a cavalier lifestyle and large family on his
twenty-five-thousand-dollar salary kept Grant saddled with debt,
but he would never cut back on his passion for fine horseflesh. He

kept twelve horses and ponies in the White House stables, plus five assorted vehicles. (At West Point he had been a laggardly student but led the class in horsemanship.) He set a record of sorts as President, becoming the only one of that august company to be arrested. Driving his spirited team along Washington's streets, he was charged with breaking the speed limit; he paid the twenty-dollar fine on the spot and commended the policeman for doing his duty, a good PR move that turned the episode into a plus.

Unlike most First Families, the Grants had lived in the capital for several years, and as the triumphant general, Grant and his Julia were lionized by Washington society. Those First Ladies who were removed from all that was familiar when they came to the White House were usually saved from isolation by their typically large nineteenth-century families.

The even dozen in the Andrew Johnsons' three-generation household provided a bulwark of support as the storm of impeachment raged around them. Washington writer Laura Carter Holloway detailed how each morning Martha Johnson Patterson—the daughter who was the stand-in First Lady—"donned a calico dress and spotless apron, to skim the milk and attend to the dairy before breakfast." (She had economically bought two cows.) With a strong sense of obligation, Martha kept all parlors and conservatories open to the public, "though many times very annoying to the inmates." (Any modern First Family would appreciate Holloway's term "inmates.") Occasionally, brazen visitors pushed their way into the White House private quarters—one such "visitor" was found napping on the Johnsons' couch.

In 1889 the Benjamin Harrisons somehow squeezed four generations—as many as eleven assorted relatives (nine of them adults) plus an Irish nursemaid—into the mansion's limited space, and even the naturally glum President seemed to loosen up with his three little grandchildren around (though a White House guard said that occurred only when photographers were present).

The John Tylers' experience was less comfortable. They arrived as a household of nine, but before Tyler's single term was over he had ousted one son for bad behavior and caused dissension in the family ranks when he added his twenty-four-year-old bride.

FRANKLIN AND ELEANOR Roosevelt, whose last two sons were away at Harvard, defied the empty-nest syndrome with a roster of permanent guests that included their divorced daughter, Anna Dall, her two youngsters—"Sistie" and "Buzzie," the nation's darlings—and several close aides of the President and First Lady. The FDR sons, when they were home, were moved about like chess pieces as guests outnumbered bedrooms. The overworked staff called it "the Grand Hotel," hardly an overstatement considering the 323 house-guests they served in the course of one year. During one such over-flow the President chuckled, "If we had one more person, we'd have to put him in the basket with [his dog] Fala." FDR would have ap-preciated the wit of young Rutherford Hayes, who joked that when he came home from college, the White House was always so full of guests he considered himself lucky to get "the soft side of the bil-liard table."

In FDR's office in the family quarters, the workday wound up in a cocktail hour, with the President in charge of the martinis. In this unusual "family" the First Lady mostly pursued her own schedule with her own friends and guests at every meal, between times receiv-ing VIPs in the state rooms (sometimes in two rooms simultane-ously) and entertaining at tea. On Sunday evenings, the staff's night off, she gathered a stimulating group for scrambled egg and bacon suppers—the only dish, she confessed, that she knew how to cook. The President rarely attended.

Eleanor loved the White House, loved it as her own home; it dimmed her memories of a bereft childhood and liberated her from her domineering mother-in-law. In the White House she could seek out thinkers and doers and not care a whit about their social status; she organized birthday parties for everybody and Virginia reels for fun. The White House that intimidated many First Ladies freed Eleanor's spirit; she reveled in its world of laughter and friends and purpose.

In her own way Helen "Nellie" Taft, for whom society would always be the capitalized noun, also reveled in the President's House; her family made themselves at home in the formal state parlors as well as the second-floor quarters. Even the East Room was put to personal use—the President, who loved to dance despite his awesome girth, organized a dance group with a dozen friends, and once a month they met to waltz and two-step to popular tunes on records.

In 1877 seventeen-year-old Helen Herron had been a guest, with her parents, at President and Mrs. Rutherford Hayes's two-day jubilation marking their silver wedding anniversary, which featured a reenactment of the ceremony in the East Room. The event must have impressed young Helen, for when her own silver anniversary came around in 1911, she was the First Lady and out to outdo any party ever given in the White House. "I am not going to try to remember how many invitations we issued," she wrote airily in her memoirs. "I know there were four or five thousand people present and that a more brilliant throng was never gathered in this country." Another fifteen thousand ogled from the other side of the fence.

The White House, completely outlined in white lights, must have looked like early Disney World, and every tree and bush was ablaze with tiny colored lights. The President and First Lady received guests under an electric sign flashing 1886–1911. Unlike Lucy Hayes, who had banned gifts, Nellie Taft definitely did not. "We were almost buried in silver," she wrote. "We couldn't help it; it was our twenty-fifth anniversary and we had to celebrate it." Everpractical Nellie collected a trove of silver to take care of all the brides she would ever know.

JOURNALIST DAVID S. BARRY, thinking back over the eleven presidencies he had observed between 1875 and 1915, concluded, "The

White House is the catalyst of a happy family life." Luci Johnson agrees. Having grown up in a family where "politics was the family farm" and LBJ was constantly tending his acres, she says, "I saw more of my parents in the White House than I'd ever seen before. I think I was sixteen years old before I ever sat down at a table with just the four of us." That is the plus side of being a privileged prisoner; the President has more time with his family. Leaving the White House for a personal evening becomes a production; reality conspires to keep Presidents at home, an unanticipated bonus for children brought up with the working nights of parents in politics.

Still, Presidents, along with their families, often chafe under the strictures of life in the Executive Mansion. "The White House is not a home," Lyndon Johnson grumped to *Life* columnist Hugh Sidey early in his presidency. "It's someplace I go when I've finished work. . . . There are a thousand conversations going on right under my bedroom with the tourists going through, and I hear every one of them. . . . And when I go up to take a nap in the afternoon, there's Lady Bird and Laurance Rockefeller and eighty women in the next room talking about how the daffodils are doing on Pennsylvania Avenue."

After years of quiet observation of this tight little world, Head Butler Alonzo Fields likened a President's situation to that of "a servant who has to 'live-in'—he begins to feel more and more cramped; the 'edge' is taken off his home life and he feels like a prisoner."

But from another point of view, life in the White House frees the First Lady from the tasks of actually running a home. She does not have to shop or cook or clean, run errands or address Christmas cards; there is always someone to press her dress, to shine his shoes, to handle social details even for their personal entertaining.

Settling into the Executive Mansion, President Reagan rated it "an eight-star hotel," and Nancy Reagan enjoyed the pampering. But at the end of their first month there, the new First Lady got a rude awakening: their personal food bill was presented to her. "Nobody had told us," she wrote in her memoirs, "that the President and his wife are charged for every [nonofficial] meal, as well as for such incidentals as dry cleaning and toothpaste." The White House is more like that eight-star hotel than she had realized; the staff keep track of every hamburger or glass of wine served to a family member or personal guest. No wonder Mamie and Bess had the kitchen save "even a tablespoon" of leftovers or that FDR was as frugal as Coolidge when it came to scrutinizing the bills. (The difference was that the hospitable FDRs dipped deep into their own pockets, while Coolidge socked away a comfortable estate from his seventy-five-thousand-dollar White House salary and expenses.) Margaret Truman wrote that, even with her mother's penny-pinching, "they had exactly $4,200 left from the President's supposedly munificent salary." In the nineteenth century, when the President's salary was low and White House families were large, many Presidents had reason to fret that the White House was leading them to the poorhouse; it was not until Warren Harding's Roaring Twenties that Congress agreed to foot the bills for official entertaining. Still, most of them had fought hard for the chance to get there.

If Nancy was shocked by their food bills, consider the tab that must have been handed to Helen Taft. Though she would not be charged for state dinners stretching over multiple courses (never quite equaling the Grants' record of twenty-nine), the President's intake was prodigious. His wife badgered him to diet for the sake of his appearance, his health, and his habit of falling asleep between courses, but when she was away, the President broke her rules like a kid sneaking two desserts.

Reporter Barry, well acquainted with the President, observed that "Mr. Taft was not even physically comfortable in the White House." At one point his doctor limited his lunch to "one apple, sliced thin," an order the President obviously disregarded. Housekeeper Elizabeth Jaffray recorded in her diary, "The President weighs 332 pounds and tells me with a great laugh that he is going on a diet but that 'things are in a sad state of affairs when a man can't even call his gizzard his own.'"

Soon a specially constructed bathtub was required because the President kept getting stuck in the standard tub, and it was hardly dignified to have to call for help. A huge vessel looking something like a sarcophagus was produced, large enough to hold four workmen sitting down, which might be thought apocryphal had it not been captured for a doubting posterity by a White House photographer. There is no record of the disposition of the tub when the very trim President Wilson took over; it would have made a fine fish pool.

THOUGH THE WHITE House is beautiful and the staff solicitous, there comes a time when a President feels like a derby-winning Thoroughbred coddled in a state-of-the-art stall or, in Truman's simile, a prisoner in the jail yard. Up until the latter years of the nineteenth century, there was not so much as a fence to prevent the public from roaming freely on the White House grounds. Since it was finally declared private, the South Lawn has served as both a buffer zone from the public and a playground for the families, and the games played there have reflected the changes in America's recreation and the enthusiasms of the individual Presidents.

Racing may be the sport of kings, but in the twentieth century golf became the sport of Presidents. The passion began with William Howard Taft, a haystack of a man whose backswing must have been a marvel to see. He enjoyed himself enormously—as he did everything enormously. Woodrow Wilson, his successor, better known for his cerebral interests, also played golf (not well) and rode horseback (not elegantly). Warren Harding, an avid golfer, would lay an old carpet on the ground and drive dozens of golf balls down the South Lawn, which Laddie Boy, his beloved Airedale, joyously retrieved. Dwight Eisenhower, solicitous of his putting green—a private gift—outside the Oval Office, had the gardeners sweep away the dew in the mornings and declared war on squirrels who buried nuts on the velvet expanse. (They were trapped and let loose in Rock

Creek Park.) The marks of Ike's cleats on the parquet floor of the Oval Office have been retained as an intimate touch of history.

Don Van Natta, Jr., in his book on president golfers, *First Off the Tee,* maintains that a President's golf game reflects his style of leadership. "Taft," he says, "would rather play than work"—he grumbled when he had to cut short a game to deal with national problems. Both Bush Presidents play what Van Natta calls "speed golf"—"it's more important to get around fast rather than best." John Kennedy was the best golfer of them all, but he kept it quiet because Ike had been sharply criticized for spending so much time on the exclusive Burning Tree Club links. After eighteen holes with Clinton, Van Natta reported, "He turned in a score of 82—but he took 200 swings."

While golf at a private club connotes a selectivity that is generally not compatible with politics, competing in an Everyman event also carries a downside: a President needs to win, preferably without breaking a sweat. Jimmy Carter was a combative athlete who would call a foot fault on his tennis opponent and larded his team with well-muscled Secret Service agents in the flacks versus hacks (i.e., White House staff versus press) softball games in Plains, but he erred in pitting himself against serious runners on a hot day. Exhausted, he had to drop out, and though his intent was positive, his image was negative. Even Jerry Ford, a University of Michigan football star and assistant football coach at Yale, was the target of endless jokes when his errant shots endangered the gallery at benefit golf tournaments.

Over the years the White House grounds have reflected the sports of the times. In 1877 Rutherford Hayes brought croquet to the South Lawn, cheering "hard-fought games with mallet and ball." Theodore Roosevelt, an evangelist of exercise, put in the first White House tennis courts, after a battle with the stingy Congress; Coolidge took up riding—on a mechanical horse in his bedroom. Harry Truman pitched horseshoes, a down-home game that was also a favorite of the first President Bush.

Throughout the years sturdy footwear and abundant energy have fueled Presidents' exercise—walking, jogging, running. Truman, an early-morning walker, led hardy newsmen on brisk forays, fielding questions as they went; Carter, Clinton, and both Bushes have been runners. Sadly, security concerns now limit the President to the track around his own backyard.

President Hoover played neither golf nor tennis nor horseshoes—games had no place in his no-fun soul. But in the interest of improvement, of giving his cabinet needed exercise, he organized medicine-ball mornings, which involved lobbing an eight-pound ball over a ten-foot net. Play began at 7:30 A.M., well before office hours, and all invitees could be counted on to attend— when the President says "Play ball!" they play ball. The activity evidently was not strenuous—the players wore suits and ties—though on one occasion the President was knocked down, hit in the face by Supreme Court Justice Harlan Stone (whose position, fortunately, was for life). There appears to have been no scoring; the real purpose, one suspects, was to score points with the President. Hoover cushioned the hardship with a breakfast of sorts, served under the towering magnolia tree planted by Andrew Jackson. On cold mornings blankets were supplied. It was quite a sight, Head Butler Fields recalled: "They ate wrapped up in blankets like so many Indians." It would have seemed simpler to move them inside for the meager toast and grapefruit, but Hoover, his record suggests, was not one to change course.

HARRY TRUMAN, WHO never lost his Missouri skepticism, famously advised, "If you want a friend in Washington, get a dog." It was good advice. From the earliest days, the President's House has accommodated pets—not to would be punitive to the family isolated in its grandeur. Traphes Bryant, who managed several administrations of White House dogs, witnessed the special role they played: "Pets help fill the emptiness." They also help the presidential image; Clinton's cat, Socks, and dog, Buddy, received more than three hundred thousand letters and e-mails. They weren't too diligent about responding—still, it made a nice book.

George W. Bush's late lamented Spotty rates a footnote in White House history: like his master, only the second son to follow his father into the White House, Spotty was the only second-generation White House pet, the offspring of former First Lady Barbara Bush's celebrated book-writing spaniel, Millie. Barney, the Bushes' saucy Scottish terrier, then took over as king of the White House, tucked under the President's arm as the two pals step out of the helicopter, starring in a family Christmas video and filming the scene with his ankle-angle "Barney-cam." The President, says Laura, calls him "the son I never had." After the victory in 2004, a newcomer arrived, a fluffy white counterpoint to Barney, Miss Beazley.

The FDRs, as they did with friends, had his-and-her dogs. Eleanor's dog, Major, made news when he bit the venerable Hattie Caraway, the first woman senator, at a White House party. Bess Furman of the Associated Press reported the incident, and when Bess was next in the White House, Major got even by nipping her nose. Franklin warned Major that one more bite and he—the President— personally would exile him to the Roosevelts' Hyde Park estate, whereupon Major, perhaps longing for those six hundred Hudson River acres, promptly bit two more guests, one of whom was the prime minister of Canada, Mackenzie King. Wars have been fought over less.

White House servants observed that President Coolidge, who scarcely spoke to his own guests, preferred animals to people. He astonished guests at an official dinner when he poured his coffee into a saucer, added cream and sugar, and placed it on the floor for his dog. The socially inept President turned to nonjudgmental pets for friendship; with them he wasn't required to make conversation.

In both the Teddy Roosevelt and Coolidge tenures, the public seemed to feel that boys in the President's House needed animals, and so the gifts came: a wallaby from Australia, a baby bear, lion cubs (quickly presented to the zoo)—the menageries in both households were variegated. Coolidge's favorite was Rebecca, a mannerly raccoon; he built her a little house of her own, but she preferred to wander at will in the President's House, startling the unwary.

The Coolidge staff used the President's fondness for animals for political ends: they trained a parrot as their lobbyist, teaching him to shriek, "What about the appropriation?" And Grace's handsome white collie, Rob Roy, achieved lasting fame, posing with her for the portrait that is a perennial favorite in the White House First Ladies collection.

John Quincy Adams could surely claim the most exotic animal— but it was more a state visitor than a pet. A White House guest might be forgiven for gasping upon entering the East Room; there, reposing in a large tub, was an alligator. It was a gift to the visiting Marquis de Lafayette, who in turn gave it to his friend, President Adams, who accommodated it with aplomb and a bathtub. In the annals of White House pets, this creature (soon displaced) remains unique, but the French hero had found a solution to what to give the friend who has everything.

OVER THE COURSE of his twenty-eight years (1941–1969) in the Usher's Office, the hub of the Executive Mansion, J. B. West witnessed profound changes in the White House. "More Secret Service, more police, less freedom of movement for the First Families, more people to manage—we were becoming more of an institution than an official residence." And that was long before security became the nation's primary concern.

The White House is a magical place that provides its families every possible perk, except one: the freedom to roam outside its walls dressed any old way, to do everything or nothing, and the liberty to mess around in the kitchen, putter in the basement, put their feet on the couch. The restraints may account for the enthusiasm both Reagan and the younger Bush have shown in clearing brush and chopping wood on their ranches, or Lyndon Johnson in taking the wheel of his big car to drive visitors around his Texas spread.

With thirty years of Washington experience, steeped in politics and surrounded by friends, the gregarious LBJs arrived at the White House with an advantage shared by both Bush families and, in the nineteenth century, the Grants and the Garfields. It makes a great difference, especially for children who have gone to school in Washington and have friends of their own; it is difficult to form easy, comfortable friendships once your home address is 1600 Pennsylvania Avenue.

Despite the mansion's pleasures, said Lady Bird Johnson, "there comes a time when the walls of the White House close in on just us four." To break free, the Johnsons headed for Stonewall, Texas, to the ranch that had been in the family for three generations, first staked out by Sam Ealy Johnson, the grandfather who drove cattle up the Chisholm Trail to Abilene.

Escaping the elegant yet somewhat claustrophobic Executive Mansion, most Presidents are drawn to their roots, to familiar hearths and extended family. And that pull of home only grows stronger with the years, even for the Franklin Roosevelts, who, in their never-equaled span of twelve years, had made the White House truly their own. Yet in 1944, in the depths of World War II, as the visibly failing Roosevelt agreed to run for a fourth term, he lamented, "All that is within me cries to go back to my home on the Hudson River." Within the year he would be buried in his beloved home's rose garden, completing the remarkable story that began the day he was born there, sixty-three years before.

Seeking family and freedom, the Kennedys headed for the familiar compound in Hyannisport, the George H. W. Bushes to theirs in Kennebunkport. The Bush compound on the Maine coast is now in its sixth generation, but George and Laura Bush prefer their brand-new ranch at Crawford, Texas, which they have imbued with White House luster, entertaining heads of state there and billing the ranch as more desirable than the White House. It "gives us the chance to have very normal, much more intimate conversations with them," Laura explains.

Harry Truman, along with the sanctuary of Bess's family home in Independence, Missouri, found liberation on the water—not in a JFK racing centerboard sloop or a Bush cigarette boat but in the luxurious 244-foot presidential yacht, *Williamsburg*. "It's just wonderful," he rejoiced. "In two minutes I'm away from everything." Sometimes he and Bess cruised the Potomac, savoring the river's tranquil beauty, returning greetings of other boaters; other times Truman brought pals aboard for nights of poker—"my favorite form of paper work"—with free-flowing drinks and inside-politics talk.

In one of the many reversals by an incoming President, Eisenhower, not a man of the sea, retired the *Williamsburg* as an economy measure, and though Kennedy would have liked to have it, he didn't dare; he settled for Ike's much smaller yacht, renaming it the *Honey Fitz*, for his maternal grandfather, Boston's first Irish American mayor. Ike and Mamie chose to renovate a two-hundred-year-old farmhouse at Gettysburg for their weekend retreat—the first house the nomad army couple ever owned.

The Theodore Roosevelts packed up their brood for long summer holidays (the government, in those days, was a part-time enterprise) at their true home, Sagamore Hill, on Long Island Sound's Oyster Bay. Years later, as he lay dying there in 1919, Roosevelt murmured to his wife, "I wonder if you will ever know how I love Sagamore Hill." Sharing that love, Edith Roosevelt continued to live there, immersed in memories of happier times before three of their sons perished in two world wars, until her death almost thirty years later.

For TR, a man who seemed to be moving even when he was sitting still, a getaway closer to Washington was imperative. He found a plain three-room farm cottage on five remote Virginia acres, with no amenities and no grace. In the spirit of his early days roughing it in the West, he did most of the cooking, for better or worse. How the serene and ladylike Edith felt about Pine Knot is not certain, but she understood that the White House alone could not contain this husband of hers, who lived life at full gallop.

The Virginia mountains also attracted Herbert Hoover, who with his wealth built (and gave to the White House) a rustic fishing camp on the Rapidan River. Forbiddingly formal to guests as well as staff in the White House, the President thawed at his getaway. According to a weekend guest, socialite Vera Bloom, "He would hold a small trusted group enthralled around a roaring fireplace, with vivid tales of his experiences as an engineer all over the world." Yet he refused to acknowledge, even with a smile, the little cluster at the White House gate hoping for a glimpse of the President—he was afraid it might be seen as a cheap bid for popularity, Vera Bloom suggested. If so, he had pathetically little grasp of communication, which surely contributed to his failure as President.

It took German submarines to bring about a safe weekend retreat for FDR. With America's entry into World War II, German subs began prowling East Coast waters—the very waters the sea-loving President often cruised. The Secret Service, insisting that he relax inland, renovated a bare-bones recreational facility built by the Civilian Conservation Corps, a Depression project in Maryland's Catoctin Mountains; continually improved, it has served all eleven Presidents since. FDR called it Shangri-la; Ike renamed it Camp David, for his grandson. The name was fixed in history in 1978 when President Carter brokered the Camp David Peace Accords between Middle East antagonists Anwar Sadat of Egypt and Menachem Begin of Israel. Dwight David Eisenhower II, his grandfather's biographer as well as namesake, acknowledges, "It's kind of fun to have something named for you."

Now only a quick helicopter hop away, the mini-resort covers a hundred forested acres at an invigorating eighteen-hundred-foot altitude, with a navy staff to provide all the cosseting of the White House with none of the public intrusions; it adapts to everything from cabinet brainstorming and summit meetings to multiple sports and easy evenings around its huge stone fireplace—and to the second wedding of the Bushes' daughter, Doro.

Ironically, Jackie Kennedy, who had rushed to rent and then, in 1963, build in the tony Virginia hunt country, belatedly discovered the delights of Camp David. "If only I'd realized how nice Camp David really is," she said ruefully to Chief Usher West, "I'd never have rented Glen Ora or built Wexford." President Kennedy confided to West that he did not like their new Virginia house and continued to use Camp David frequently; Jackie had actually raised the question of selling Wexford. Then the issue became moot.

Imagine the discomfort of nineteenth-century First Families—White House captives sweltering in those layers and layers of clothes in the swampy capital's suffocating heat, in a mansion dependent on the occasional breeze. (Central air-conditioning was installed with the rebuilding in 1948.) But there was a presidential getaway—not in the same league with Camp David but a godsend, a gingerbread-and-gables house at the Soldiers' Home in Washington. Only three miles north of the Executive Mansion, its three hundred protected acres were considerably higher and breezier, and it boasted a postcard view of the Capitol dome.

From their second year the Lincolns spent long summers in the presidential cottage set aside for his family, a total of thirteen months. In his book *Lincoln's Sanctuary*, historian Matthew Pinsker provides the fullest picture of their life at Soldiers' Home. The President, a commuter ahead of his time, was driven to the White House along with other carriage traffic, though a cavalry escort was ordered as Washington seethed with reports of plots against him. (A bullet actually landed very close to him, but he dismissed it as a soldier misfiring his rifle.) After his day at the office, Lincoln would return to the retreat, shuck his boots, slip into carpet slippers, and sit on the front porch, enjoying the evening with Mary, playing checkers with Tad.

It was a familiar, Illinois kind of life, a normal life the White House did not offer. Tad, nine, befriended the soldiers in the military guard, who made him a "third lieutenant"; saddling up his pony, he would trot down to their camp and join the chow line, occasionally accompanied by his father, who liked talking with his troops over a plate of beans. Lincoln found escape from the pressures of the White House and moments of personal peace at this place that was more like himself, plainer and more remote. On April 13, 1865, a fine spring day, he rode out to Soldier's Home to discuss the dates the family would arrive for the summer. The following day he took Mary for a carriage drive, a pleasure they could share more often now that the war was behind him. That evening, humoring Mary, he took her to see a light comedy at Ford's Theater.

The youngest First Lady at twenty-one, Frances Folsom married Grover Cleveland, forty-nine, in the only White House wedding of a President. During his second term Frances gave birth to the first presidential child ever born in the White House. *Library of Congress*

Julia Gardiner, twenty-four, married President John Tyler, a widower thirty years her senior. She referred to her eight months as a First Lady as "my reign," received guests in a "queenlike state," and liked being addressed as "Mrs. Presidentress." *Library of Congress*

Jacqueline and John F. Kennedy—the youngest elected President and the third youngest First Lady at thirty-one—introduced three-year-old Caroline and baby John to the White House, the home that would be theirs for a thousand unforgettable days. *John F. Kennedy Library*

April 10, 1865, a day after the South surrendered, a serene President Lincoln is photographed with his mischievous Tad—the last sitting before the assassination four days later. *Library of Congress*

And you meet such interesting people—Amy Carter, a bright eleven-year-old, shakes hands with Soviet Union leader Leonid Brezhnev. She also met Pope John Paul II when he visited the White House. *Jimmy Carter Library*

The Clintons celebrate Chelsea's fifteenth birthday in the homey nook they created in the family kitchen. *Clinton Presidential Library*

Like any seventeen-year-old, Susan Ford washes her car in the family's driveway—but at the White House there's always a long lens lurking. *Gerald R. Ford Library*

As the media crowd around her father's limousine, Jenna Bush, twenty-two, expresses the view shared by most children living in the White House. *AP/WideWorld Photo*

Archie and Quentin Roosevelt, White House scamps, report for duty with the White House police. Their imagination for mischief was unlimited. *Library of Congress*

Against the din of Vietnam protests, Lyndon Johnson finds joy in meeting his first grandchild, Lyndon, Luci's son. *Lyndon Baines Johnson Library*

David and Barbara Anne Eisenhower race their tricycles at their second home—their grandparents' fun White House. *National Park Service*

A four-generation family of eleven somehow fitted into the Benjamin Harrison White House. Russell, his father's secretary, holds his daughter's hand; "Mr. Whiskers" pulls the McKee grandchildren, Benjamin and Mary. *Library of Congress*

Above: Caroline and John take Macaroni to visit the President at his office—he thought it wiser to meet alfresco. *John F. Kennedy Library*

Left: Grace Coolidge cuddles Rebecca, their beloved raccoon. Her husband preferred animals to people. *Library of Congress*

Below left: Harry Truman advised, "If you want a friend in Washington, buy a dog." President Ford confers with Liberty in the Oval Office. *Gerald R. Ford Library*

Below: LBJ and Yuki sing a duet for Ambassador David Bruce. Yuki, the President's favorite pet, had been abandoned at a Texas gas station when Luci rescued him for a life of White House luxury. *Lyndon Baines Johnson Library*

President Taft tots up his score after a round of golf. His course handicap: about 330—pounds. *Library of Congress*

A tough competitor, Jimmy Carter takes a break back home in Plains and shows no mercy to the White House press corps on the softball field. *Jimmy Carter Library*

A personal putting green outside the Oval office was not enough for avid golfer Eisenhower—but frequent games at a restricted club drew criticism. *U.S. Navy Photo*

"Poker," said President Truman, "is my favorite paper work." On a break from official papers he takes advisers Clark Clifford (right) and George Allen (left) plus a couple of other old pals for a cruise on the presidential yacht. *Harry S. Truman Library*

Accomplished equestrienne Jackie Kennedy shows her horse how to take a fence. *John F. Kennedy Library*

President Hoover escapes his burdens with a rod and reel in a trout stream— in solitude and in a high-collar shirt and tie. *Herbert Hoover Presidential Library*

The first President Bush takes daughter Doro for a spin on the snowbound slopes of Camp David, the rustic adjunct to the White House. *George Bush Presidential Library*

Newly widowed President Wilson, flouting the gossips, fell in love and—against all advice—married Edith Galt in 1915. Later, concealing his stroke, she acted as virtual president. *Woodrow Wilson House*

After a White House romance, Lynda Johnson married military aide Charles Robb in a Christmastide wedding in the East Room. Afterward they began a life in politics together. *Lyndon Baines Johnson Library*

Tricia Nixon added a page to White House social history when she defied the threat of rain and married Edward Cox in the only wedding ever to take place in the Rose Garden. *National Archives*

Nancy Reagan's svelte gowns were the epitome of 1980s chic—and they got her into hot water. *Ronald Reagan Library*

The fashion of the times are mirrored by First Ladies' gowns. In 1861 spendthrift Mary Todd Lincoln indulged in extravagant hoop skirts and a wreath of flowers for her hair. *Library of Congress*

In 1902 madcap Alice Roosevelt, the first White House debutante, beguiled the nation with her frothy gowns and haute society guest lists that made front-page news. *Library of Congress*

Beautiful young Frances Cleveland displays the hour-glass figure that every woman yearned for in 1886. She and Jackie Kennedy are the only presidents' widows to remarry. *Library of Congress*

Eisenhower's state occasions were American regal with the President in white tie, the First Lady in diamonds, and the Marine Band in red tunics. *Dwight D. Eisenhower Library*

Legendary Pearl Bailey has a new accompanist for her White House gig: President Nixon playing the East Room piano. He also serenaded Duke Ellington with "Happy Birthday." *National Archives*

President Clinton revels in a Walter Mitty day playing his sax with jazz greats at the 1993 White House Jazz Festival. They gave him high marks. *Clinton Presidential Library*

Suddenly, the White House dance floor clears. Diana, Princess of Wales, the world's most glamorous woman in 1985, dances with Hollywood superstar John Travolta at the Reagans' dinner in her honor. Who would dare compete? *Ronald Reagan Library*

At the historic event honoring all living American Nobel Prize laureates, President Kennedy talks with author Pearl Buck and the First Lady with poet Robert Frost. JFK declared it "the most extraordinary collection of talent, of human knowledge, ever gathered at the White House—with the possible exception of when Thomas Jefferson dined alone." *John F. Kennedy Library*

Hillary Cinton serves Russian leader Boris Yeltsin champagne; he serves her floating moose lips. *Clinton Presidential Library*

Nancy Reagan, attacked for borrowing designer clothes, is the hit of the Gridiron Club's annual show lampooning herself in a thrift shop outfit. *Ronald Reagan Library*

Susan Ford, forgoing blue jeans for sophistication, is reassured by her father that she will be a fine substitute for her mother who is recuperating from cancer surgery. *Gerald R. Ford Library*

From the time Prince Albert (later King Edward VII) visited President Buchanan in 1860, British royalty have created a stir as White House guests. Eleanor Roosevelt accompanies Queen Elizabeth (later the beloved Queen Mother) from Union Station in 1939. *Library of Congress*

A young Tricia Nixon was paired with Prince Charles when he visited the White House in 1970 with his sister, Princess Anne. Washington matchmakers were disappointed that no romance ensued. *National Archives*

In 1957 the Eisenhowers greet the youthful monarch, Queen Elizabeth II, early in her long reign. *Dwight D. Eisenhower Library*

Long before the civil rights movement, Eleanor Roosevelt championed black Americans. In 1939, defying public censure, she invited world-famous Marian Anderson to sing at the White House and presented her to the British King and Queen. Later, at the United Nations, Eleanor was named humanitarian to the world. *National Archives*

Lady Bird Johnson and Interior Secretary Stewart Udall spot an eagle on one of her many trips to awaken Americans to their nation's beauty—and to protect it. *Lyndon Baines Johnson Library*

Spurning officials in 1909, Helen Taft set the tradition that the new First Lady rides with her husband in the inaugural parade. Later it was her vision that wreathed Washington in its world-famous cherry trees. *Library of Congress*

Pat Nixon blossomed on her solo foreign
trips to Africa and South America—in
Liberia she goes native. And the crowds
loved her. *National Archives*

In South Africa Nelson Mandela shows
Hillary and Chelsea Clinton the cell that
was his solitary prison for twenty-seven years.
Clinton Presidential Library

Rosalynn and Jimmy Carter, true partners, harness their post–White House prestige to fight
disease and hunger in the third world. Here, they work in Ghana. *The Carter Center*

After her breast cancer operation, Betty Ford became a public advocate for early detection measures. *Gerald R. Ford Library*

To encourage children to read, former librarian Laura Bush and the Library of Congress stage an annual book festival on Washington's National Mall. In the wake of Hurricane Katrina, the First Lady organized a foundation that has raised more than $1 million to restore the 206 school libraries destroyed across the Gulf Coast. *Library of Congress*

A glimpse of history: the six living First Ladies in Washington in 1994. *Left to right:* Nancy Reagan, Lady Bird Johnson, Hillary Clinton, Rosalynn Carter, Betty Ford, Barbara Bush. Each left a lasting imprint on the White House—and it on her. *Clinton Presidential Library*

seven

The Mirror of the Times

O N THE EVENING of November 8, 2000, the White House was at its finest, celebrating its two hundredth anniversary, a historic milestone, with an event that was itself historic. Nine of the guests felt a touch of nostalgia as they entered the familiar halls— they were the unprecedented assemblage of Presidents and First Ladies who had come to salute the mansion they'd been privileged to occupy. Former President Jimmy Carter captured the essence of the gathering with an account of his recent meeting with an international group. "I asked, 'What do you think is the most famous building in the world? One said 'The Taj Mahal,' another said 'Buckingham Palace.' The rest of them said 'The White House.'"

From its beginning the White House has been synonymous with America. It has been the mirror of the nation's new ways and old customs; it reflects what American ladies are wearing, the food they serve, and how an American parlor looks; it conveys the shifts in manners and mores, the changing role of American wives. A history lover, leafing through old White House images, sees one era morph into the next and the next. From artists' illustrations and Gilbert Stuart portraits to crackled daguerreotypes and Mathew Brady's new-era photographs, from stiff-necked poses to action in color, they are the nation's family album.

As the planning began in 1790, President Washington was adamant: the presidency must be housed in an edifice that would demonstrate a new nation ready to take its place among the capitals of the world. It must be "imposing," he insisted. "In size, form, and elegance it shou'd look beyond the present day." Long before the first spade of earth was turned, Washington anticipated enlarging the President's House when the country's "wealth, population and importance shall stand upon much higher ground than they do at present." Unyielding in his vision, he headed off a proposal to put both the President and the Congress in the same structure, deflected a suggestion for renting a presidential residence, and pressed a pinchpenny Congress to come up with the funds for his "imposing" plans.

Along with fellow Virginian Thomas Jefferson, Washington had been the force behind the controversial decision to build the permanent Federal City so far south, in a swampy, mosquito-infested, malaria-ridden stretch of nothing, bogged down in mud in winter, choked by dust in summer. A man could go along Pennsylvania Avenue with his gun and bag partridge for his table. (Nowadays President Washington would be accused of conflict of interest in pushing this location, since it would surely increase the value of Mount Vernon, his plantation fourteen miles downriver. He even nudged the site further down than the boundary by Congress.)

It fell to the reluctant John Adams to shepherd the move of all 136 government employees from Philadelphia to the new "city." The witty, elitist Gouverneur Morris, a Founding Father who arrived in Washington as a senator, expressed the general attitude with his flippant report from the new capital: "We want nothing here but houses, kitchens, well-informed men, amiable women, and other trifles of the kind." But he added, presciently, "It is the very best city in the world for a *future* residence."

Moving into the President's House, Adams assiduously followed the tone set by the Washingtons in the first presidential residence, in New York. Staunch revolutionary that he was, Adams harbored a fondness for the trappings of the royal courts of Europe. Like Washington, Adams, in powdered wig and velvet, greeted guests with a formal bow from the waist (an opposition newspaper ridiculed him as a "mock monarch"). Like Martha, Abigail received guests while seated regally on a raised platform. But even Adams felt that Washington's ornate carriage drawn by six white horses was a bit kingly and limited his own carriage to two, elegantly named Caesar and Cleopatra.

More troubling was Adams's argument for a title grander than mere "President," though at least he opposed "King" and "His Highness." It is forgotten, these two hundred years later, that the chief executive's title was not yet chiseled in granite. Deriding the plainness of "President," a then-unprecedented title for the leader of a sovereign nation, Adams sniffed, "What will the common people of foreign countries, what will the sailors and soldiers say, 'George Washington, President of the United States?' They will despise him to all eternity." Fortunately, Adams the political thinker was more reliable than Adams the social prognosticator.

Jefferson, the patrician Virginia plantation owner, might seem more likely than the New England Puritan to have clung to social deference, but the erudite third President brought to the Executive Mansion a fierce commitment to a new citizen style. He preferred to be addressed as plain "Mr. Jefferson," and he greeted all comers not with a formal European-style bow but with a democratic handshake.

At his inauguration Jefferson signaled his rejection of pomp, choosing to walk the few hundred yards from his boardinghouse to the Capitol and afterward walking back to Conrad and McCunn's to sup with the usual friends around the communal table. Adams, who scorned the walk as lacking in dignity, had left the city on the 4 A.M. stagecoach. The bone-rattling fifty-six-hour journey to Boston (eased by overnight stops and the occasional Holland gin laced with sugarhouse molasses) was preferable to witnessing any part of his former Vice President's triumph.

The White House—the nation—was blessed that the larger-than-life George Washington was followed by two Presidents with such deep commitment to the new concept of a republic. (Consider the upheavals in contemporary newly independent nations.) Adams and Jefferson each brought strengths that both reflected and shaped the changing nation. Adams, demonstrating that the esteemed General Washington was no fluke among the upstart nation's leaders, sought to establish the White House as a stable power center qualified to take its place among the royal courts of Europe. Jefferson, pointing the way toward a more egalitarian republic, sought to establish citizen power and the primacy of the individual in the new Executive Mansion and the new nation. The definition of who was or was not a citizen lay in wait decades down history's road.

No one gave a thought to a title for Abigail, other than "Mrs. Adams." Some referred to her as "Madame President," "Mrs. President," or, more monstrously, "Presidentress." At first, Martha Washington had been saluted as "Lady Washington," a title she found quite suitable, but the new America's leaders wisely rejected it as too royalist. Eighty-nine years would pass and sixteen women would preside over the President's House before the President's wife had a title of her own, awarded not by law but by a newspaperwoman. "First Lady" initially appeared in 1871 as a description of Julia Grant; the title was bestowed more formally on her successor, Lucy Hayes, by Mary Clemmer Ames, a journalist covering the 1877 inauguration. Though it sounds a bit archaic—and most of its latter-day holders have disliked it—the title serves to recognize the President's wife as an individual in her own right, not merely an appendage of her husband.

GEORGE WASHINGTON, A trained surveyor, was a hands-on overseer of the President's House from its inception; with the city planner, Pierre Charles L'Enfant of Paris, he pinpointed its precise site and kept an eye on the competition to select the best design. James Hoban, an Irish-born architect who had settled in Charleston, won the honors and a five-hundred-dollar prize. His plan for the structure, which would serve as both home and workplace for the chief executive, appears to have been inspired largely by a classic eighteenth-century showplace, the grand estate of the Duke of Leinster. Leinster House still stands proudly in Dublin, the seat of both houses of the Irish Republic's parliament.

The house for the American President, however, would be a modest, even austere, version. Bare and boxy, it sat undressed on a site that could boast of little but a vista sweeping down to the magnificent Potomac River; in early drawings the house is scarcely recognizable as today's majestic Executive Mansion.

The cornerstone was finally laid on October 13, 1792, after which the official party repaired to a Georgetown inn to celebrate, lifting their tankards to everything they could think of—to the fifteen states, to liberty, to the fair daughters of America. Some reported sixteen rounds of toasts, others said thirty-two—who was counting? At any rate, there were enough that the masons forgot to record where they had placed the cornerstone, and to this day it has never been found.

Years passed before trees and shrubbery provided their gentling embrace, three decades before the hospitable porticoes were added, with the stately columns that would became the building's signature. As late as 1841, after nine Presidents had lived in the White House, celebrated British novelist Charles Dickens commented on its still-raw appearance. The garden, he wrote, "is pretty and agreeable," but the President's House "conveys that uncomfortable air of having been made yesterday."

Eight years in construction, the President's House fulfilled Washington's demand: Hoban's classic mansion, designed with Washington and Jefferson looking over his shoulder, was indeed imposing. Still, as Abigail Adams made clear in her candid letters, it was by no means finished to meet the December 1800 deadline set by Congress. Some things never change.

The President's House was to be the epicenter of official and social life in a nation finding its own character. The Capitol (not completed until 1825) was a daytime workplace for legislators more concerned with their own states than with the Federal City. The "city" boasted a few fine homes built by those of affluence and influence seeking closest proximity to power, and, facing the square across from the President's House, four modest brick buildings accommodating the entire executive branch of government. Otherwise, there was only a cluster of shabby little dwellings along the riverbank and workers' (many of them slaves) shacks everywhere, evidence of the city's coming transformation into a world capital.

Georgetown, established fifty years earlier, was the center of what commerce there was with its busy port and streets cluttered with six-horse Conestoga wagons loaded with the bounty of Maryland farms. Abigail Adams, who had no other place to shop, called Georgetown "the very dirtyest Hole I ever saw for a place of any trade or respectability of inhabitants." She would disbelieve today's chic enclave, which proudly retained its federal charm as the city expanded around it.

ANY HISTORY OF the President's House and its families must include the singular tale of First Lady Dolley Madison's dedication under fire. In 1814, as the capital city and the President's House were beginning to fulfill their promise, America was once again at war with England. After two years of battles on land and sea, King George III's forces had advanced to Bladensburg, a Maryland town just east of the District of Columbia line.

Dolley, almost alone in the White House, could hear the boom of guns in the skirmish. President Madison and his generals were on the battle scene, in despair that the ill-prepared American militiamen had fled before even one man was lost. Her husband dispatched urgent instructions to her: gather personal valuables and his important papers and leave immediately, as he was forced to do.

She could see smoke rising from the fires as the invaders burned their way through the city, advancing relentlessly toward the Capitol and the President's House. Dolley proved more courageous than the militiamen. Disregarding personal danger, in an astonishing hour-by-hour chronicle to her sister Lucy in Kentucky, she described the two days when war overtook the White House. "Two messengers, covered with dust, come to bid me fly," she scribbled. "I am determined not to go until I see Mr. Madison safe, and he can accompany me." The hundred men detailed to guard the President's House had vanished; "disaffection stalks around us."

On the morning of the collapse of hope, she continued the historic letter that is held today by the Library of Congress. "Since sunrise I have been turning my spyglass in every direction," but there was no sign of her husband. "Alas, I can descry only groups of military wandering in all directions, as if there was a lack of arms, or of spirit to fight for their own firesides." She had filled a carriage with trunks stuffed with her husband's papers, and miraculously a wagon was found for the remaining documents and the family silver.

As the day wore on she wrote Lucy, "I am still here within sound of the cannon!" (Her husband by then was at a safe distance.) Then—like a war correspondent broadcasting from battle—she detailed as it was actually taking place the event that has given her permanent rank in the pantheon of American heroes. "Our kind friend, Mr. Carrol[1], has come to hasten my departure, and is in a very bad humor with me because I insist on waiting until the large picture of Gen. Washington is secured, and it requires to be unscrewed from the wall, This process was found too tedious for these perilous moments—I have ordered the frame to be broken and the canvas taken out; it is done—and the precious portrait placed in the hands of two gentlemen of New York for safekeeping. And now, dear sister, I must leave this house, or the retreating army will make me prisoner in it, by filling up the road. When I shall again write to you, or where I shall be tomorrow, I cannot tell!"

Her letter is breathtaking in its unruffled command as the enemy, unchecked, sent the Capitol and every government building up in flames. At the last possible moment Dolley fled. British troops broke down the White House doors, dined on the supper that had been laid out for the President's staff, and then, brandishing their torches, they roamed the mansion, piling up furniture for a bonfire, setting every room ablaze. Throughout that night, it was said, the glow of the flames made Washington as bright as midday.

Dolley's travails were the stuff of novels. After a night in an army camp just beyond Georgetown, she crossed the Potomac to relative safety in Virginia. (The President was hiding out in the woods.) Seeking a room in a tavern, the First Lady was turned away by the barkeep's surly wife, furious that her husband had been conscripted into the army. That evening a god-sent August hurricane doused the fires destroying the city as the British raced back to their ships in the Baltimore harbor. Dolley, in disguise and on her own, slipped into a farm wagon, then, revealing her identity, crossed the river on an army ammunition barge and made her way on foot to a house belonging to her other sister.

Long after the war, a contrite British officer wrote, "I was horrified at the order to burn the President's House. I shall never forget seeing [everything] go up in flames. Our sailors were artists at the work." The White House stood open to the sky, a skeleton of blackened walls. Its rooms were destroyed, but its spirit—embodied by the undaunted Dolley—remained intact. A more important symbol than ever, it was rebuilt and three years later was once again alive. And today the Washington portrait is the only item that survives from the original White House.

IF WE THINK of Washington in the 1800s as a place of formality and morality, of ruffled shirts and minuets, early journalist Perley Poore sets the record straight. The capital he wrote about "resembled, in recklessness and extravagance, the spirit of the English seventeenth century, rather than the dignified caste of the nineteenth." Occasionally, violence erupted on the floor of Congress; dueling was accepted in defending honor or settling a grudge. In one memorable incident in 1853 a California congressman, irritated by a waiter's lackadaisical service, pulled out a pistol and shot him dead between the soup and the fish.

Even before the government had unpacked, the lobbyists were in place in their town houses, using their well-appointed parlors to persuade, offering good food for goodwill. In Poore's view the "most adroit" lobbyists were women, proper widows or former officials' daughters—and some not so proper. "Who can blame a congressman," the journalist mused, "for leaving the bad cooking of his boarding house, to walk into the parlor web which the cunning spider-lobbyist weaves for him?" Once again, after two hundred years the fundamentals are little changed, except that food is secondary to campaign contributions.

The White House and its residents have always changed with the times—sometimes with enthusiasm, sometimes not. In 1842 Samuel Morse asked Congress for twenty-five thousand dollars to prove the value of his revolutionary invention, the telegraph. It was ridiculed as "a matter of merriment," and one congressman scoffed that part of the money should be used to send a telegram to the moon. A shift of three votes in Congress would have consigned Morse's invention into oblivion. Even Perley Poore, reporting the debate, could not appreciate the irony of the scornful jibe about the moon. President Tyler, however, foresaw its value, and within the year the telegraph was a major tool at the White House.

By 1848 it seemed high time to bring gaslights into family

quarters—over the protest of First Lady Sarah Polk, who insisted that the large chandelier in the entrance hall be lit by candles. When the new gaslights were proudly turned on at the first reception, Sarah reveled in an I-told-you-so moment: the gaslights flickered, died, and all went dark—except her brightly candlelit reception hall. One modernization that no one opposed was central heating, of a sort, which came in 1853.

It seems impossible, in the mansion now blessed with twelve and a half baths, that Dolley Madison had only a wooden bathtub, filled by buckets—a nicety that Andrew Jackson, three Presidents later, removed as "undemocratic," though he later agreed to have water piped in. Not until 1851 was a bathtub equipped with water from taps, and President Fillmore was roundly chastised for this "monarchial luxury." At that same time the White House kitchen was updated with the first real cookstove, a thing of drafts and pulleys that disconcerted the cook, who preferred old-fashioned iron pots and a big fireplace. The President, refusing to have progress derailed by a cook, put aside matters of state to go to the Patent Office for a lesson in working the new invention. The stove was converted, but was the cook? There is no report.

A generation later, in 1877, Rutherford Hayes, fascinated by Alexander Graham Bell's new device, the telephone, took part in a personal experiment with the inventor. As the President listened, thirteen miles away from Bell, a reporter witnessed "a gradually increasing smile wreathe his lips and wonder shone in his eyes." Straightaway the White House boasted its own telephone with its own number: 1. Decades later Lyndon Johnson not only had telephones in his bathroom and on a leg of the family dining table, he had one installed on a tree outside his LBJ house.

In 1891 electricity finally came to the White House—and yet another First Lady, Caroline Harrison, had reservations about a newfangled amenity. She and her husband, Benjamin, refused to touch the switches, fearing lethal shock. They would use only the old gas mode, summoning the young White House electrician, Irwin

Hoover, to flick on the electric lights. Having proved so handy at turning the lights on and off, "Ike" Hoover stayed on for forty-two years, in charge of the entire mansion.

NINETEENTH-CENTURY FIRST Ladies might well have said of the President's House, It's a nice place to visit, but I wouldn't want to live there—because it really wasn't what anyone would call home. Throughout the Executive Mansion's first century the Presidents' wives gamely struggled to create a family life in a near-impossible situation. The second floor of the White House was divided between the offices of the President and his cabinet on the east end and the family quarters on the west, separated only by a half-glass partition and complaints.

Though each sector had its own stairway, all comers used the same front entrance. Inside, the reception hall was a veritable public square, thronged with flotsam and jetsam seeking jobs, asking favors, pleading a cause, polishing connections. In those days the best way to land a much coveted government job—where the pay was regular and no manual labor was involved—was to badger the President himself, so the stairway to the executive offices was crowded with supplicants waiting their turn. The scene was hardly compatible with a family home.

An irritated President Polk grumbled in his diary that "neither ice nor fire [could deter] the herd of lazy, worthless people" who wasted his time, yet he was obliged by custom to put up with them. Lincoln, suffering a mild case of smallpox, with his unfailing humor could joke about the favor seekers. Show them in, he said—"Now I have something I can give to everybody."

In those conditions a First Lady was required to bring up her children literally in the lap of government. "It's like living over the store," protested the patrician Edith Roosevelt. As the executive branch grew ever larger, a separate residence for the President's family was suggested, with the White House reserved for offices and state occa-sions. The proposal got nowhere: the Founding Fathers had envi-sioned the White House as the heart of the nation, and the family's place was at that heart. No further discussion.

It took a family that overflowed its allotted half of the second floor to force a solution. In 1902 Theodore Roosevelt, acknowledg-ing that it was impossible to place limits on his large and rambunc-tious brood of six, was the catalyst for the solution: tear down several large greenhouses and scruffy maintenance shacks and build an inte-grated addition for offices. It would become the West Wing.

In the same major renovation, a balancing East Wing was added, following the general plan envisaged by Jefferson. The East Wing came to include the ever-expanding offices of the First Ladies. It was Edith Roosevelt, the prototypical nineteenth-century upper-class wife, who ushered in the expanded the role of the First Lady as a public figure. She was first to have a staff assistant—paid fourteen hundred dollars a year by the government—and first to formalize the running of the White House, hiring a manager (the chief usher), who answered to her.

Going into its second hundred years, the White House had finally given the First Family the entire second floor to call their own; at last they could truly feel at home. At that point Theodore Roosevelt made what seemed a small change. He had his official stationery im-printed "The White House," and as simply as that—no committee, no discussion, no legislation—the Executive Mansion was given the name that would come to be synonymous with American power around the world.

IN THE NEW century the country was changing fast, and the vener-able old mansion turned chameleon. If a new generation wanted to listen to the radio—even dance to it—or play games, wear slacks, eat hamburgers, entertain young men without the ever-present Secret

Service, the White House boasted just the place for fun living. It was the solarium. Added in 1927, when the mansion's roof was raised to add eighteen rooms on a full third floor, the new room was at the very top of the White House. It quickly became the place where a President can put his feet up and the younger generation can be young; there are no antiques; the casual look changes with each family's lifestyle. But its most exciting feature never changes—its huge windows offer a breathtaking view of the great monuments bathed in floodlights, the National Mall alive with activity, and the river gleaming in the distance.

James Hoban could not have imagined such a room in his formal plan. But then he couldn't have imagined the President taking over its kitchen to cook his special beef stews for old friends, as Ike did, or Caroline Kennedy and her twenty little schoolmates coloring and singing in their sunny kindergarten, or the LBJ daughters keeping their fiancés out of sight. "It was like a family room," Susan Ford explains. "We used it a lot for playing cards, listening to music, and watching TV. I took several dates up there just to hang out." The solarium's informality, its view out with no view in, provides a free and easy escape from the museum house downstairs. It is, in Lady Bird Johnson's apt term, "the citadel of the young."

Among those museum rooms, the East Room, the largest in the mansion, equals the solarium as a chameleon space. Visualized by Hoban as "the great audience room" in the royal manner, it has been the setting for five grand weddings of White House daughters and bivouacked rough Union soldiers. It has accommodated Teddy Roosevelt's jujitsu sessions with his Japanese trainer, debutante dances, serious conferences, Abigail Adams's laundry drying on a line, children's parties, world-famous performers, one christening, too many funerals, a reception for P. T. Barnum's famous midgets, Tom Thumb and his bride, Nobel Prize laureates, opera, ballet, Lawrence Welk, a Chinese wrestling match, and presidential press conferences. No one can say that the President's House doesn't bend to meet its families' interests.

Activities in the smaller red, blue, and green parlors tend to be conventional meet, greet, and eat affairs. Mary Todd Lincoln, however, used the Red Room for a strange purpose: séances with a spiritualist medium who promised to communicate with her dead sons. To placate his wife, unbalanced by sorrow, the deeply skeptical President Lincoln attended one session—but remained skeptical.

OVER THE YEARS the President's House has been repainted (no question about the color), repaired, refurbished, restored, reconstructed; there have been additions, subtractions, and innovations; it has undergone four major renovations and has come close to being torn down to be replaced with an entirely new structure, proposals that fortunately were rebuffed. But in 1948 something had to be done: the old house was about to collapse around the Truman family.

At a reception for the DAR in the Blue Room, Bess Truman was startled to hear the crystal prisms of the huge chandeliers tinkling. In the family quarters overhead, Alonzo Fields, a giant of a man, was serving President Truman's lunch when, as Truman described it, "the floor sagged like a ship at sea." Across the corridor, daughter Margaret's piano was sinking into the floorboards. No question, the grand old building was beyond repair; it had to be replaced.

Fortuitously, this crisis was met by the chief executive who was the most knowledgeable historian of all of the Presidents. A special commission offered three alternatives: demolish completely and build anew; demolish and replicate using new material; or preserve the outer walls, gut the interior, and rebuild from the inside. Truman supported the third option.

He was adamant that every possible piece of wood, every architectural detail, be returned to its former position. Each piece was meticulously labeled and stored; what could not be salvaged was replicated. The vast undertaking required 660 tons of concrete and steel and took almost four years; even the rebuilding in the wake of the British burning in 1814 was of less scope. Without Truman's advocacy, the President's House, with all the history in its bones, could have become just another postwar building.

During the reconstruction, the Trumans moved into Blair House, diagonally across Pennsylvania Avenue (later, it was officially designated the President's guesthouse). The dislocation posed no hardship for Bess and Harry, who were much happier in the less grand, family-size substitute—and Bess was relieved of many of the social functions she heartily disliked.

FROM THE BRITISH invasion of 1814 until today, the guns of war, blazing on the other side of the world or the hemisphere or in what are now Washington suburbs, have echoed throughout the White House again and again, rumbling through the lives of its families.

In the seconds it took a radio voice to deliver the shattering bulletin on Sunday, December 7, 1941, the White House changed. Japan's attack on Pearl Harbor brought machine guns to the mansion, and in its deepest basement a shelter, one hundred by fifty feet, was installed, encased in four feet of concrete, sealed by heavy steel doors, and equipped with its own power plant, water supply, and advanced communications system. Troops stood guard twenty-four hours a day; airspace over the White House area was closed. Throughout the night guards patrolled the lawns, and on one occasion two men with guns, demanding to see the President, were arrested. All windows were covered with heavy black cloth, but when the Secret Service wanted to seal them with cement blocks and the army proposed painting the White House black, Eleanor led a revolt of all who lived or worked in the mansion.

Washingtonians hurrying to work must have done a double take: surely that was not the First Lady striding up Connecticut Avenue, unaccompanied. It was indeed Eleanor Roosevelt on her way to her job at the Office of Civil Defense, wearing a uniform of sorts, a blue-gray coverall of her own design. Her effort to broaden OCD to include her ideas for social change brought a barrage of criticism and the decision that she could do more good for the war effort in other ways. She worked ceaselessly, visiting servicemen abroad, entertaining them at the White House, and comforting the wounded in their hospital beds.

The President and First Lady showed no fear, though a gas mask was attached to FDR's cumbersome wheelchair, to be within reach at all times. Taking heed of threats against him, his food was given special scrutiny and kept in locked boxes. All tradesmen were investigated, and the White House bought its own delivery truck, manned by two Secret Service agents—one stayed in the vehicle, while the other monitored every item of food as it was handled. Like those of every American family, the Roosevelts' meals were circumscribed by ration coupons; Eleanor took hers with her when she traveled, and White House guests were expected to do the same.

More than any home in the country, the White House had gone to war, and it would never return to the relaxed ways of the past. The First World War had been different; the grim technology of death was not yet sophisticated. In Woodrow Wilson's wartime White House, the family set an example with meatless days, wheatless days, and a vegetable garden; the First Lady's sewing machine clacked away as she made pajamas and hospital shirts for the Red Cross, and the President's daughter Margaret cajoled him into letting her go to France to sing for the troops. In his own eccentric war effort, President Wilson brought a flock of sheep to the White House grounds—they not only trimmed the grass, their ninety-eight pounds of wool raised $49,333 at auction for the Red Cross.

Three generations later war once again affected the family in the White House—but this time the family was the enemy as anti–Vietnam War protesters filled Lafayette Square across the avenue. Even after forty years, Luci Baines Johnson cannot forget the chant that permeated the White House and kept her and her sister, Lynda, sleepless at night: "Both of our husbands were in Vietnam— and the last thing we heard before we went to bed, and sometimes the first thing in the morning, was people shouting, 'Hey, hey, LBJ, how many kids did you kill today?' People think you're isolated from public opinion in the White House, but we certainly were not."

IT WAS THE Civil War that brought conflict into the capital city and the White House itself. It had seemed a wise decision in 1790 to locate the new Federal City at the join of the fledgling nation's two parts, the North and the South, but then came the time when the line no longer united the nation but divided it. By 1860 Washington was a bifurcated city, southern by geography and inclination, yet the President's House was at its heart and the heart of the Union cause.

The muddled President Buchanan, meeting with a stern delegation from the border slave states, wept openly at his indecision. Beset on all sides, he was eager to leave the White House, which he had mostly enjoyed as a splendid setting for his lavish entertaining. As the clouds of war darkened, journalist Perley Poore observed the scene around him: "It was as in Paris just before the revolution of 1830, when [former French foreign minister] Talleyrand said to Louis Philippe at a Palais Royal ball: 'We are dancing on a volcano.'"

From White House parlors to family dinner tables, Washington was riven by bitter division among friends, even families. A prominent artist who mixed in society but not in politics, Marietta Minnigerode Andrews, described the times: "In every house, in every phase of life and every class of society, the air was full of threats and insults, of open affront and secret slander. There was fear everywhere, fear of assassination, fear of spies."

On his way to Washington, President-elect Lincoln, warned that he was targeted for assassination, sent his wife and children from Baltimore to Washington on a different train, and wore a shawl and traveling cap to alter his appearance. One week after his inauguration, the Confederate States of America adopted a formal constitution; one month later, Fort Sumter, in the Charleston harbor, was attacked. The war that sliced through the soul of the nation had begun. Elizabeth Todd Grimsley, Mary Todd Lincoln's cousin, shivered at the mood she felt during a visit with the Lincolns in the White House: "There is a feeling of danger lurking in the air." To the dismay of the Lincolns, guards were stationed inside the White House.

Washington, with its southern tilt, was riddled with Confederate spies, and a number of them were discovered to be "ladies of wealth and fashion" who hid army secrets "in the meshes of unsuspected crinolines." Proper gentlemen smuggled Union Army plans in the linings of "honest-looking coats," and military maps were stolen from the War Department.

In the very shadow of the White House, and perhaps within its parlors, those artful ladies, part Mata Hari, part Walter Mitty, were working to overthrow Lincoln's government. The prominent Mrs. Rose O. H. Greenhow boasted that information she had gleaned from a New England senator had determined the outcome of the battle of Bull Run. She did indeed receive a letter thanking her on behalf of Confederate President Jefferson Davis: "We rely upon you for further information. The Confederacy owes you a debt."

But Mrs. Greenhow got her comeuppance when her fine house was used to detain female spies, with military guards stationed at tents in her garden. Those ladies were not game playing; they gathered information and transported funds—one was carrying seventy-five hundred dollars when she was arrested. Mrs. Greenhow was jailed in the Old Capitol Prison, then "deported" to Richmond, capital of the Confederacy. President Jefferson Davis dispatched her to Europe to plead the Confederate cause in such high-level meetings as a personal appointment with the Emperor of France, Napoleon III.

The Union cause countered with its own web of lady spies operating in Richmond, the Confederate capital, directed by the well-connected Elizabeth Van Lew. This spy came in out of the cold much better off than Mrs. Greenhow—President Grant made Van Lew postmaster of Richmond, one of the highest federal positions open to a woman in those days.

eight

The White House à la Mode

JACQUELINE KENNEDY, CURIOUS about what might be found in the unseen White House, pulled on a pair of old jeans and a vintage sweater and prowled through the dusty storage rooms, the nooks and crannies of her new home, and its warehouse. Her search for furniture from earlier times was a letdown. "I was disappointed to find hardly anything of the past in the house," she lamented. "Hardly anything before 1902." She did find one lost treasure—a pier table from James Monroe's fabled French purchases had wound up in storage, sprayed with gold radiator paint.

The new First Lady's commitment to the White House was deep and genuine; she arrived with a vision and determination. Early in the first year she told *Life* columnist Hugh Sidey about her plans: "The minute I knew that Jack was going to run for President I knew that the White House would be one of my main projects, if he won."

Revealing a quirky bit of fantasy, she confided, "When I first moved into the White House, I thought, I wish I could be married to Thomas Jefferson because he would know best what should be done to it. But then I thought no, Presidents' wives have an obligation to contribute something, so this will be the thing I will work hardest at myself. How could I help wanting to do it? I don't know . . . is it a reverence for beauty or history? I guess both. I've always cared."

Then she shared a private thought that makes the heart ache now, knowing as we do how it all turned out: "When I think about my son and how to make him turn out like his father, I think of Jack's great sense of history."

Wasting no time, she put her vision into action. "I want to make this a grand house," she explained to Chief Usher West, whose long experience as White House manager made him her invaluable assistant. Guided by a prominent New York decorator, they set to work. "Mamie pink" disappeared; dark walls were lightened to white; Victorian mirrors were declared "hideous—off to the dungeons with them!" The Grandma Moses painting banished by Ike found a place on Caroline's rosebud-papered walls; twenty of George Catlin's famous Indian portraits were retrieved from the Smithsonian.

Then the serious work began. "Everything in the White House must have a reason for being there," she insisted. "It would be sacrilege merely to 'redecorate' it—a word I hate. It must be restored—and that has nothing to do with redecoration. That is a question of scholarship." For expertise she established a Fine Arts Committee and, for more practical purposes, the White House Historical Association "to enhance appreciation and enjoyment of the Executive Mansion"—and raise funds for acquisitions. Of course she had solid advice in all of this, but the concept and the personal commitment were hers.

By the end of two whirlwind weeks Jackie had used up the entire fifty-thousand-dollar appropriation for redoing the White House family quarters; not a nickel was left for the state rooms. "Never mind," she reassured Mr. West, "we're going to find some way to get real antiques into this house." And she did. As a result of special attention lavished by the glamorous Jackie, wealthy Americans quite happily dug deep to purchase costly antiques for the President's House or gave from their own collections. As a result of Jackie's vision, and the continued efforts of other First Ladies, the White House state rooms are enhanced by more great period pieces today than they had in the actual period. For American visitors there is almost a sense of ownership—Pat Nixon used to remind tourists, "This is *your* house"—and pride that it is the only residence of a head of state that receives gifts and funds from individual citizens.

In considering the success of Jackie's White House restoration, pity the plight of John Tyler, the tenth President. He spent his entire term in a running battle with Congress, which not only denied him any funds for much needed refurbishing but made him pay for lights and fuel in the Executive Mansion that housed his offices as well as his home.

There was a reason—deplorable—that Jackie's explorations produced so little old furniture. In the frequent refurbishing of the White House over the many decades, the furniture of the outgoing families was often relegated to storage, sold, or chucked out. Chester A. Arthur, the twenty-first President, was the most egregious culprit. Moving into the White House following the assassination of President Garfield in 1881, Arthur, a wealthy widower prominent on the Manhattan social scene, took one look at his new home and declared it "a badly kept barracks. . . . I will not live in a house looking this way. If Congress does not make an appropriation, I will have it done and pay for it out of my own pocket." With that, he took up residence in the home of a friend until the White House was redone to his posh standards.

As for the furnishings, he decreed, "Out with it all!" Twenty-four wagonloads of everything, some of it threadbare junk but some objects of historical interest, were carted away for auction. Five thousand eager buyers snapped up such items as lace curtains from the state reception rooms, the globe Nellie Grant had used to study geography, brass cuspidors—everything.

Arthur was of the "best that money can buy" school—vintage French wines, a "nobby" wardrobe, and a carriage that Smithsonian expert Herbert Collins deemed "the swankiest turnout in White House history." To do over the White House, Arthur turned to Louis Comfort Tiffany, the most famous decorator of the period. (Today the Metropolitan Museum devotes a room to his work.) The showpiece was a huge stained-glass screen, one of Tiffany's grandest designs, installed in the entrance hall, setting the tone for the haute-Victorian makeover. Gilt was applied everywhere with the abandon of mosquito spray. A few years later Arthur's splendid Tiffany screen, the height of fashion in the 1880s, was lost to posterity in the same cycle of dismantle and replace.

Had there been a thoughtful First Lady to oversee the transition, perhaps she would have saved some of the earlier pieces from the auction block. White House castoffs, even though not important antiques, are sought by the Smithsonian Institution, "the nation's attic," and other museums. No matter how bad the taste, how overwrought the decor, White House furnishings characterize their time; they show us how First Families lived, how America lived.

Twelve Presidents later, history buff Harry Truman was still sore at Arthur: "He sold a pair of Lincoln's pants and some of Jackson's furniture. The whole thing only brought in about four or five thousand dollars, and there were some things that were just invaluable."

FROM THE EARLIEST days new First Ladies moving into the White House have experienced, in nearly equal parts, a sense of wonder and a wave of dismay at the shabby state of things, especially following a two-term family. The run-down condition is not surprising in a house that—until terrorism imposed stringent restrictions—has played host to many hundreds of dinner guests, thousands at larger events, and up to two million tourists in a single year.

In the FDR years, Eleanor's untrained and high-handed housekeeper, Henrietta Nesbitt, decided "it was high time the White House had some new furniture." With that, Mrs. Nesbitt got rid of the remainder of President Monroe's exquisite French pieces. "The old Monroe furniture didn't match [her new draperies]," she wrote, quite proudly, "so we sent it to the National Museum."

When the White House was rebuilt following its destruction by the British in 1814, there was little money to refurnish it. President Monroe eased the problem by selling to the government the fine Louis XVI pieces he had purchased while minister to France—some of it bought at the auction of Marie-Antoinette's belongings. At the bargain price of $9,071.22$\frac{1}{2}$—the records are exquisitely specific—the government got the best of the deal. That incident alone demonstrates the wisdom of "Jackie's law" preventing passing fads and quirky personal tastes from taking over the historic salons. Years later the White House reclaimed two priceless Monroe chairs.

The mansion's tolerance was severely stretched to accommodate Theodore Roosevelt's decor for the State Dining Room, newly enlarged in the extensive renovations that were made when the new wing for the President's offices and his staff was added. The big-game hunter's prized fully-antlered moose head was installed in the place of honor over the fireplace, where it glowered down at VIP diners. Lesser trophies were dispersed around the room. The Tafts lived with them—TR had bestowed the presidency on Taft—but the Woodrow Wilsons promptly dispatched the herd of baleful animals to a corral in the Smithsonian and hung tamer paintings in their place.

Some First Ladies have not fallen under the spell of the White House, most notably Julia Grant, who denounced it as "not suitable for a President's residence" when she proposed using the Executive Mansion only for official events. But the next First Lady, Lucy Hayes, was charmed by its classic beauty. "No matter what they build," she exclaimed to a visitor, "they will never build any more rooms like these," and, like Jackie Kennedy years later, she rescued old china and furniture from the "lumber rooms." Julia Grant and Lucy Hayes were of the same generation, from the same Midwest, of the same political persuasion, yet they saw the White House through different eyes; Jackie and Lucy, different by every measure, shared the magic of history all around them.

In 1873 Julia redid the mansion to her taste, in the florid "Mississippi riverboat" excess that was *le dernier cri* at the moment. The East Room, as described by White House historian Amy La Follette Jensen, was "an overdecorated horror in the 'Greek style.'" Gilded wallpaper, new furniture of ebony and gold, white and gilt woodwork in great profusion, "completed the destruction of the once stately room." A vogue for the gaudy had taken over; it was a low point in high style for the White House. But Julia loved her new look and wept at leaving the President's House after eight golden years.

Julia would have been pleased that 130 years later another First Lady, Laura Bush, poking around in White House storage, brought out a set of the Grants' furniture. With a closer look at it, she decided "that's sort of Victorian-looking—it's not that attractive. And my husband said, 'No, I want Grant's furniture in my office.' He thought it was really terrific." There it has stayed. She and her husband have both learned from the lives of their predecessors, she said. "We are very interested in what life was like for them." Luckily for the White House, modern First Ladies have been more responsive to its spell than their nineteenth-century predecessors—but then the White House has also become more responsive to its residents.

While the historic self of the White House dominates its families, it indulges their preferences on the family floor. A First Lady can choose colors and rearrange furniture as she wishes, banishing things that don't appeal to her and rediscovering others. If Ida McKinley wanted to hang horseshoes in her bedroom for luck, no one objected. If Eleanor Roosevelt wanted to cover every tabletop with mementos and every wall with family pictures, only the maids groaned—silently. If Lou Hoover wanted Chinese furnishings in the historic American house, in they came.

The most famous piece of furniture on the second floor is the oversized rosewood bed that Mary Todd Lincoln purchased for her husband—though White House historians doubt that he ever used it. Over the years it has been moved around like a chess piece; no President has wanted the massive bed as his own, but none would put it in storage. Lou Hoover, who shuffled furniture around through her entire four years, repeatedly moved the bed in and out of storage, never certain what to do with it. Its wanderings led Harry Truman to suggest, "Now I know why Lincoln's ghost walks around up here at night—he's just looking for his bed." For himself, Truman replaced FDR's hospital bed with an antique canopied four-poster; Bess retained Eleanor's single bed for her own, smaller room; but there was, it seems, more communication between the Trumans' rooms.

The largest room in the family quarters, the beautiful yellow drawing room that echoes the oval shape of the Blue Room on the floor below, has been used over the years for the Millard Fillmores' library, the Rutherford Hayeses' hymn fests, Franklin Roosevelt's study, and all kinds of personal entertaining. Most important, it is now the setting in which the President and First Lady receive foreign leaders, exchange gifts, and break the ice before a state dinner. In her diary, Lady Bird Johnson mused, "I like to recall that the famous men and women who came before us knew the same curious mingling of public and private activities."

Like much of the President's House, the Yellow Room is too formal for relaxing. For comfortable family evenings, the best the White House can provide is the west end of the broad corridor that bisects the structure. It was the scholarly Lou Henry Hoover—she spoke four languages, knew two more, and was a connoisseur of Asian arts—who first visualized the space as an informal living area, adding bookshelves, artwork from their years of living abroad and traveling the world, and her beloved canaries. Though she contributed much to the White House and her West Hall Sitting Room has done much to make the White House livable for its families, she gets little credit—the Hoover administration never caught the public fancy. "This is the only place that is not sacrosanct," Lady Bird said of it when the Johnsons moved in. "This is our very own." So she immediately hung a portrait of a famous Texan—LBJ's mentor, legendary Speaker of the House Sam Rayburn.

Speaking for all contemporary First Families, Lady Bird was grateful to Jackie for bending the White House to suit today's families. After a disastrous dinner with a three-year-old in what she called the "cavernous" family dining room on the state floor, Jackie knew that wouldn't do. She persuaded the White House watchdogs to convert the corner suite that had been Margaret Truman's into a more realistic family dining room—labeled "the President's Dining Room"—with its own kitchen. At last, after 160 years, the White House provided its families a fully contained home on the second floor, removed from the ghosts of history that inhabit the state rooms.

The Clintons later took family informality a step further, putting a small table and comfortable wicker chairs in the kitchen for the three of them. (Still, when Hillary wanted to cook a favorite dish for Chelsea, she had to override the kitchen staff—in the White House nobody relinquishes domain easily.) Such sticklers for propriety as the Monroes and the Hoovers would surely raise their eyebrows at such casual dining, but the President's House was once again reflecting a changing America.

NEVER WAS THE President's House in a more deplorable state than when the Andrew Johnsons moved in, two months after the tragedy of President Lincoln's assassination. The distraught Mary Todd Lincoln had stayed on, a hollow, disintegrating widow with no concern, understandably, for housekeeping. Union soldiers assigned to the White House in the war years had treated it like a barracks, leaving the stately East Room filthy and vermin-infested; rugs, draperies, and upholstery were beyond saving. Taking over, Martha Johnson

Patterson, the President's daughter and a senator's wife, personally wielded mop and pail to redeem the dilapidated mansion.

The Trumans, going through their new home for the first time, were similarly dismayed. The second and third floors showed the ravages of twelve years of wear and tear by the Roosevelts' large, three-generation family, resident aides, constant guests, and multiple dogs—and were not improved by Eleanor's lack of interest in domesticity. Margaret Truman never forgot her first impression: "I remember crying myself to sleep on my first night in the place. It all looked so shabby." The family quarters were "a mess. . . . It looked like a third rate boarding house." The long-serving Lillian Parks deplored the worn and dog-damaged carpets, draperies held together by tape, and rats that were "regular visitors." Soon Margaret's spacious suite, with her grand piano installed in the sitting room, was transformed into dream quarters for a twenty-one-year-old college senior and her pals. The White House has "good bones"; it has adjusted to families of every size and inclination.

But even the President's House has its problems. Every First Lady has struggled to keep the mansion warm. Abigail Adams burned cords of wood in multiple fireplaces trying make the place fit for guests, and even tough old Andy Jackson, who had fought and won great battles, growled, "Hell itself couldn't warm that corner." The Trumans resorted to electric heaters to stop their shivering, and two administrations later Jackie Kennedy fretted, "Surely, the greatest brains of army engineering can figure out how to have this heated like a normal rattletrap house!" And it's still cold.

THEY STAND ON tiptoe, crane their necks, strain for a better look. The First Lady has entered the room. While they applaud, they examine her outfit; while she speaks, they take mental notes to pass on to friends. Bright blue again to set off her white hair . . . the new hairstyle is not as flattering as last week's . . . her hem isn't straight . . . how can she stay so thin and eat all those White House dinners? The next day, though she won't admit it, the First Lady studies the newspaper photographs or reviews a clip from the evening news to make her own judgments: the new designer suit tames the hips, the legs look good in the higher heels, but that handbag is as awkward as the Queen's. For both audience and object, the First Lady's appearance tends, at first, to overwhelm what she has to say. "I always hate the stories that talk about how I look," Laura Bush says with a sigh. But it is inescapable—with the White House comes scrutiny.

The constant appraisal and rude opinions lead to one undeniably positive result: every First Lady leaves the White House more poised, more fashionable, than when she entered. Despite the added stress and heavier responsibilities than she has ever known, the attention—and flattery—can make a First Lady feel prettier and more confident. The White House demands more of her, and if she is strong, it enhances her sense of self.

Laura Bush is only the latest example of a First Lady who quickly rose to the challenge. Her first appearance in Washington after the drawn-out election of 2000 exposed her to the kind of gratuitous critique all First Ladies dread: her lavender plaid outfit was declared unbecoming, not a good color, not well cut. It was never seen in public again; in its place came brighter colors, better lines— and a red Oscar de la Renta suit. Laura was still Laura, but a sleekened Laura. Back in 1909, Nellie Taft was another quick study. Scarcely a month after she became First Lady, the President's closest aide, Major Archibald Butt, privately observed, "She really looks ten years younger since she entered the White House, and I think she has become more gracious and kinder toward all the world." President Taft was pleased. "I love to see her well dressed," he confided to Major Butt. "The only promise I extracted from her was that she would not economize in dressing . . . and as you may have discovered, economy is her prevailing mood."

Even the most self-assured new First Lady may be intimidated by the judgmental press and public. Hillary Clinton's constant fussing with her hairstyle was evidence that even an issues-oriented lawyer with years of experience in the Arkansas governor's mansion still worried about how she looked in the President's House. Eleanor Roosevelt, minimally concerned about clothes during her twelve years in the White House, later counseled new First Ladies, "You will feel that you are no longer clothing yourself; you are dressing a public monument."

From the earliest days in the still-unfinished White House, everyone asked the burning question: What was the First Lady wearing? Abigail Adams, dispirited as she was in her brief tenure, dutifully carried out a full social program while making the great barren "castle" livable. In her daily life she dressed in the modest New England style, but on official White House occasions she blossomed, as fashionable as a duchess in rich brocades befitting the status of the American President's wife. Her necklines were cut low—discreetly filled in with an organdy fichu—and like her mentor, Martha Washington, she covered her hair with a puffy white "mob cap."

Eight years later the much admired Dolley Madison moved into the President's House in a burst of style and merriment, a full turn away from her Quaker upbringing and Abigail's example. After a long apprenticeship as the widowed Jefferson's frequent hostess, she was ready to set her own pattern for the White House, using its prestige and elegance to create the political and social center for the still-developing capital. At the Madisons' first inaugural ball Dolley set the pace, resplendent in a Paris creation of buff velvet with a long train, topped off by her fashion trademark, an exotic turban of matching velvet, spiked by a bird-of-paradise plume. Her towering turbans and high heels made her—an exceptional five feet nine inches tall before adding plume and heels—stand out all the more, which was the point. For eight years the public never took their eyes off of Dolley. Her husband, "Jemmy," was a towering intellect, but at five feet four he could be, as one writer put it, "in danger of being confounded with the plebeian crowd . . . pushed and jostled about like a common citizen."

At those very early White House entertainments, Washington's two or three women reporters overlooked no detail, led by Mrs. Samuel Harrison Seaton, who reported all the gossipy bits for her husband's newspaper, the *National Intelligencer.* Describing one reception, after references to the stifling heat, she made her point bluntly: "You perhaps will not understand that I allude to the rouge which some of our fashionables had unfortunately laid on with an unsparing hand, and which assimilating with the Pearlpowder, dust and perspiration made them altogether unlovely to soul and to eye." Such inside tidbits were then served up for teatime gossip.

Rouge, it should be noted, was equated in those days with questionable ladies, and there was ongoing debate as to whether Dolley herself might occasionally add a pinch of color to those rosy cheeks. In Dolley's defense, Mrs. Seaton loosed her artillery of adjectives: "Her majesty's appearance was truly regal—dressed in a robe of pink satin, trimmed elaborately with ermine, a white velvet and satin turban, with nodding ostrich plumes . . .'tis here the woman who adorns the dress, and not the dress that beautifies the woman." The First Lady is "admired by the rich and beloved by the poor," she declared, then tossed another morsel to her readers: "You are aware that she snuffs, but in her hands the snuff-box seems only a gracious implement with which to charm." Mrs. Seaton's next invitation would be in the mail.

As the century ticked by, fashion escalated from the soft empire lines of Elizabeth Monroe and Louisa Adams to the billowing layers of petticoats, then into the skirts belled by hoops (actually cages) beloved by Mary Todd Lincoln and Julia Grant, skirts of such proportions that the simple act of walking through a White House door required careful maneuvering. Tiring of that excess, the First Ladies of the latter quarter of the century turned to excess of a new kind—the grotesquerie of bustles making the simple act of sitting down difficult and waistlines so tightly cinched by corset strings that even the simple act of breathing was in jeopardy. The construction gave the wearer the forward tilt of a ship's figurehead straining into the wind.

Though this was the peak of Victorian pecksniffery, necklines were cut provocatively low (President Grant objected, but Julia wore them anyway) and the expanse of white shoulders was tantalizing. The concept of too much of a good thing was unknown in nineteenth-century fashion—with furbelows and flounces, ruching and ruffles, paniers and trains, berthas and sashes and rosettes, an entire vocabulary of ornamentation was born. Fashionable dressmakers piled one conceit upon another for confections of such elaboration that in 1877 journalist Perley Poore cheered Lucy Hayes's "plainly arranged hair and high-necked black silk dress" and the disappearance of "party dresses cut so shamefully low in the neck as to generously display robust maturity or scraggy leanness."

The overnight change in First Ladies sometimes caused temporary consternation among the early Washington ladies who looked to the White House to set fashion. In 1843 young and stylish Julia Tyler—only twenty-four, she was President John Tyler's trophy wife—briefly electrified the scene. Eight months later, thanks to the vagaries of politics, her unelected husband lost his presidential perch and she was gone.

Glamorous Julia was replaced by Sarah Polk, serious-minded and deeply religious. Regulars at White House affairs knew from the moment they scrutinized the Polks at the inaugural ball that they would have to rethink their wardrobe. The new First Lady wore a gown of such plainness as to be a message: Julia's New York ways were gone; East Tennessee would be setting White House style, and it would be simple and modest.

Twenty years later another Tennessee family was thrown suddenly into a grieving White House. Andrew Johnson's wife and daughters, like Sarah Polk reserved and unpretentious, made no attempt to live up to Mary Todd Lincoln's dramatic New York gowns, bright-hued, low-cut—and outrageously extravagant. The Johnson ladies, ever aware of the horrendous circumstances that deposited them in the White House, chose modest dark dresses, elegantly detailed, with chaste white fluted collars at the high necklines. The effect was appropriate for a subdued White House and was much admired—though not as satisfying to the ladies of fashion.

For the first 160 years of White House couture, stylish ladies made a statement with the hats they chose: from demure mob caps to big, bold, colorful, feathered, veiled creations. Poet Edgar Lee Masters wrote of a time when ladies believed that "hats could get husbands or lose them." One guest, writing in mid-nineteenth-century overkill, described a reception at the Polk White House with an extravagance that matched the chapeaux: "Milliners sent [the ladies] forth in fit trim to challenge the rainbow for the exquisiteness and variety of colors in which they were decked, while on their heads and bosoms glittering brilliants recline like nestling glow-worms, darting forth rays of light in dazzling emulation." The writer seemed to suggest that the ladies wore colorful hats and lots of jewelry.

In the 1950s, the last decade of universal hat-wearing in America, Mamie Eisenhower's closets held two hundred of them, pert little numbers that did not obscure her signature bangs. In addition, Mamie's many dresses filled two rooms, a First Ladies' record, with Lou Hoover a close runner-up.

Never has the White House known a fashion phenomenon to equal Jacqueline Kennedy. A lustrous beauty, bewitching, enigmatic—and headstrong—at thirty-one she influenced the whole of America and upper-crust women around the world with her wardrobe of sleek, pared-down designs that bespoke unerring taste and top-drawer breeding. In 2002, forty years after the days of Camelot, crowds waited in line for two hours at New York's Metropolitan Museum of Art to see an exhibition of clothes she had worn in her brief White House years. The whims of fashion had changed many times over those decades, but the dresses, suits, and evening gowns remained chic, elegant, timeless. The late Eleanor Lambert, founder of the Best Dressed List and American fashion's most enduring exponent, made this salient point: "What people don't understand is that Jackie was not about *fashion*—Jackie was about *style*."

While Jackie was in the White House, her clothes were followed more closely than the Dow Jones index, and Mrs. America did not hesitate to tell the First Lady what she was doing right or wrong. Her mailbags bulged with admiring letters—along with letters carping about skirts cut too short, strapless gowns cut too low. Jackie tartly instructed her staff director, Letitia Baldrige: "Tell them I'm having my skirts shortened and my strapless gowns cut much lower." The esteemed Ms. Baldrige, who was experienced in both politics and diplomacy, dispatched notes to all detractors: "Mrs. Kennedy was interested to hear your opinion, and she has asked me to send you her very best wishes." The letters are probably framed for posterity by those newly minted friends of Jackie; her real replies would have been more memorable.

In the 1960 campaign Republican strategists had seen Jacqueline Kennedy as a liability. American voters, they figured, would deem the young society beauty, the product of privilege, as not a suitable First Lady, not in comparison with Pat Nixon, the spunky daughter of hard times and hard work, the wife who was always at her husband's side. But they misjudged: Pat Nixon was attractive and admirable, but Americans love superstars. Then came a Republican charge that Jackie spent ten thousand dollars a year on clothes (an even more enormous sum in those days). Again the Nixon camp misjudged: in a front-page interview with Nan Robertson of the *New York Times*, Jackie shot back, "I couldn't spend that much on clothes unless I wore sable underwear!" The political attack boomeranged—her laser comeback revised the image of Jackie the mannequin into Jackie the self-assured new-era First Lady.

IT IS NOT only First Ladies who have influenced and reflected American styles. The portraits of the Presidents, their suits, their hair, their very expressions, track the changing times. John Adams followed the style of his predecessor, George Washington, who had followed the styles set long before in the royal courts of Europe. Adams received the first guests in the new President's House wearing typical eighteenth-century apparel: powdered hair (or wig) caught by a bow in back, ruffled shirt, satin coat, black velvet knee breeches, silver buckles on his slippers.

Twenty-four years later his son, John Quincy Adams, marked the transition to a new era as the first President to wear long trousers, plain black, at his inauguration. Between breeches and trousers much had changed in the country and in the White House—a war had been fought, political parties had been formed, the White House had gone through its first rebuilding. And the age of Presidents in velvet and lace had ended.

What a President wears can convey a message. Washington and Adams signaled that the new American government was not in the hands of a bunch of rowdy revolutionaries. Jefferson, to display change from Adams's royal formality, used the White House to nurture his democratic society of Everymen, which sometimes tended to slip into studied slovenliness. (One unfriendly senator regaled his friends with the shocking fact that the President received him in a frayed brown—brown!—jacket. What would he have said generations later had he attended a meeting with a pajama-clad Lyndon Johnson?)

Most Presidents give more thought to their appearance once they settle into the White House; the place demands certain standards. Lincoln, the frontier lawyer, was always self-deprecating about his scarecrow frame and lack of polish, and the press agreed with him. When his wife, Mary—who was consumed by appearances—outfitted him with a new overcoat from Brooks Brothers, purveyors of propriety, the *New York Tribune* congratulated her in an editorial.

In the presidential fashion stakes, Chester A. Arthur is the clear winner. If a new style emerged in New York, he would have it—in a single order he had his New York tailor make twenty-five coats for various occasions. Not surprisingly, he was the first President to require a valet. When Woodrow Wilson came to the White House in a new era, his limousine was specially built with a roof high enough to accommodate his tall silk hats, a small concession to presidential panache.

Calvin Coolidge, frugal—not to say stingy—as he was, weakened when the mills of his native Vermont presented him a bolt of finest suiting. He indulged in his first custom-tailored suit and liked the result; it looked right for the master of the White House. But his gossipy Secret Service agent revealed that under his bespoke suits the President still wore huge suspenders and underwear that was "three sizes too large," accoutrements that, happily, were not on view.

Harry Truman, the purposely plain man of the people, both by nature and in contrast with the aristocratic FDR, was a President who could, and sometimes did, iron his own shirts and even wash his own underwear. At his winter retreat in Key West, Harry was wild about flashy sports shirts, to Bess's despair, but she won the decision on the red pants, a gift, banning them as not presidential. Truman, the only haberdasher to gain the White House, was probably the most clothes-conscious President of them all. In natty double-breasted suits, bow ties, and, in summer, two-tone shoes, his panama hat worn with just the right snap, he could have been a model in his own Kansas City establishment. Despite his "give-'em-hell" image, Truman was a very particular man.

In today's White House President Bush reflects his fellow Americans. In the Oval Office he dresses as conservatively as a Wall Street banker, but as soon as he sets foot on Texas soil, he climbs into the jeans and cowboy boots that feel like home, just as his father had donned golf togs and boating gear in Kennebunkport. Though the White House is the goal, the mountaintop, it's a relief to be able to look like any guy once in a while; back on his own turf, every day is casual Friday and a President can tap into his real self again.

A PRESIDENT MIRRORS the style of his times—and influences them—in the way he deals with his face. For the first seventy-two years Presidents were clean-shaven, until Abraham Lincoln arrived with a beard—to hide his ugly face, he liked to say. ("When people call me two-faced," he joked, "I ask, would I wear this face if I had another one?") After that, only one of the ten chief executives between Ulysses Grant and Woodrow Wilson was bare of beard, mustache, or mutton-chop sideburns; that was William McKinley. In those days, as the man in the White House shaved, so shaved the nation. During World War I beards vanished (the Kaiser wore a beard), and early movies made mustaches the mark of a rogue, a villain, or a comic. Facial hair, once so manly, was scrubbed from politics until the 1940s when Republican presidential candidate Thomas Dewey's little mustache became his trademark. "He looks like the little man on the wedding cake," quipped Alice Roosevelt Longworth, whose wicked wit was bipartisan. That image could have been the margin that gave Truman his upset victory in 1948—he won by a whisker.

nine

Twenty-nine Courses,
Four Thousand Hands to Shake

N<small>O MATTER HOW</small> fiercely she would guard her privacy or, in times past, how little she planned to do, every new First Lady arrives aware that like it or not, prepared or not, she has just become the nation's leading hostess. In that role she will put her personal stamp on the White House in the atmosphere she creates, the entertainment she selects, the menus and flowers she chooses, and the guest list. And everybody will feel free to comment and criticize.

White House entertaining is more than having a lovely evening in a majestic setting—Congress would not appropriate fifty thousand dollars a year (plus another twenty thousand if needed) just for the President to have parties. Breaking bread together has long symbolized friendship, which is the purpose of state dinners honoring visiting heads of state and for receptions recognizing good works and achievement. A White House invitation itself acknowledges that the recipient is a leader in some field—business, arts and entertainment, sports, politics—or at least a campaign contributor.

The use for political purposes is not surprising; from its early days the White House has been an instrument of friendly persuasion. When feasible, Lincoln invited sympathizers of both sides of the divided nation and at the war's end instructed the Marine Band to play both "Dixie" and "The Battle Hymn of the Republic" at a concert on the White House grounds. (He might be astounded to learn that a night in his bed has been used as a quid pro quo for campaign contributions.) Only Woodrow Wilson, the idealist son of a Presbyterian minister, insisted, "I will not permit my house to be used for political purposes."

A White House evening can be much more than high-level mingling. A four-star dinner washed down by the best of American wines in the very heart of American history can do much to advance a delicate diplomatic balance, and though the East Room may be a small venue, it offers a world stage to American artists. They are paid nothing; their honorarium is to be introduced by the President or First Lady, to be applauded by crowned heads and world leaders, to meet celebrities of every stripe—a heady experience for even the most renowned artists.

A guest might be treated to unplanned special moments—like the evening Van Cliburn, the Texas-born pianist whose mastery of Tchaikovsky made him a favorite in the Soviet Union, played "Moscow Nights" at a Reagan dinner and the Gorbachevs sang along. Guests that evening sensed the coming thaw in the cold war. Years earlier, in 1961, an artistic thaw occurred when legendary Spanish cellist Pablo Casals ended his almost-sixty-year boycott of the United States (and any country that recognized the fascist Spanish government) to perform for—and warmly embrace—President Kennedy. Among the guests was another legend, Alice Roosevelt Longworth, who had attended Casals's previous White House performance, in 1904, during her father's administration. And when the irrepressible Pearl Bailey cajoled Egypt's Anwar Sadat into dancing with her, it definitely warmed the visit to the Nixon White House. An awesome roster of great artists have sung, played, danced, and acted in the East Room, a tribute to the White House couples who reached out to the finest.

Lyndon Johnson's daughters would often slip into the East

Room's back row to hear world-famous performers, and guests would sometimes catch a glimpse of four-year-old Caroline Kennedy in pink pajamas and robe, perched at the top of the stairs, wide-eyed as she watched the glamorous scene below. Susan Ford, graduating from a private school for girls, hosted her senior prom there—a first in White House social annals. "To have it in the White House was unique," she says. "Everybody in the class showed up, and that never happens." Small dinners preceded the dance, and Susan, with her four best friends and their dates, dined on the presidential yacht, with the moon rising over the Potomac. Young memories don't get much better than that.

The great social events become a part of White House history, illuminating the style of the presidential couple and their times. Forty years have slipped by, and Jackie Kennedy's spectacular state dinner at Mount Vernon is still unmatched. Inspired by President de Gaulle's grand dinner for the visiting American President in the palace of Versailles's Hall of Mirrors, Jackie wangled the use of George Washington's home to honor an important new American ally, Pakistan's Ayub Khan.

Guests were ferried down the Potomac in PT boats (a reference to JFK's heroic career in World War II); mint juleps from Washington's own recipe were served; the Colonial Color Guard and fife and drum corps, in full Continental Army uniform, performed a Revolutionary War drill that was first ordered by Washington; dinner was served in a Tiffany-decorated tent overlooking the river, and the Washington National Symphony performed. (The concert was a close call—at 4 P.M., in direst emergency, an acoustical shell had to be built for the sound to be right.) Army troops served their country by spraying the entire grounds, twice, against hordes of mosquitoes. It was a historic evening—and too overwhelming a task to be repeated.

The ultimate salute to excellence was the Kennedys' dinner for Nobel Prize winners of the Western Hemisphere, an unequaled array of brilliance. In his famous toast, the President declared of the twenty-nine laureates, "This is the most extraordinary collection of talent, of human knowledge, that has ever been gathered together at the White House—with the possible exception of when Thomas Jefferson dined alone."

If that event presented an unequaled array of brilliance, Teddy Roosevelt's dinner, in 1902, for Prince Henry of Prussia takes the honors as the most elaborate stag affair ever held in the White House. The thirty-six select guests were mostly military officers, listing to port under the weight of their medals as they ate their way through seven courses, from *huîtres sur coquille* to *marrons glacés*. The decorations were fancier than a nouveau riche debutante's coming-out party, a spectacle fit for a prince, or even the German Kaiser, who happened to be his brother.

A stereopticon slide (the turn-of-the-century equivalent of a video) shows a photograph of the banquet, with text exulting in the "royal welcome to the gallant Prince." The East Room was transformed by strings of tiny electric lights suspended from the ceiling; garlands of smilax festooned the huge chandeliers; ropes of greenery, mountains of roses, masses of azaleas filled every available space. German and American flags bespoke friendship.

Punch was served in small boats flying the flag of the Kaiser's new yacht, built in a New Jersey shipyard and christened by America's princess, Alice Roosevelt. To add a German touch to the toasts, Teddy ordered beer instead of champagne; the White House, having no steins, rented them from a local restaurant. "When the Prince lifted his stein for a toast," a guest reported, "he almost choked to read printed on the bottom, 'Stolen from Gersternberg.'"

The stereopticon text praised "national hospitality warm in its greeting to royalty, without departing from republican traditions." Twelve years later Germany was at war with America's allies and then with America itself. Dinner-table diplomacy proved to be only seven courses deep.

WHAT WOULD SURELY have been the most glamorous of all White House dinners was one that never happened. The guests of honor were to be Princess Grace of Monaco and her husband, Prince Rainier, during their visit to Washington. Grace Kelly had reigned over Hollywood as its most elegant star and Rainier's Grimaldi family had reigned over the tiny principality of Monaco since 1297 (with time out for the French Revolution), making them the oldest royal dynasty in Europe. It had all the makings of a sensational evening but for one detail: Jackie Kennedy refused to have the dinner. It seems that dashing young bachelor Jack Kennedy and the ravishing Grace Kelly had long ago enjoyed a brief romance.

Jackie tuned out. Plans for a glittering state dinner were downgraded to a minimal luncheon. Letitia Baldrige, the First Lady's superb chief of staff, later attributed this pout to "a bit of jealousy, perhaps"—Jackie was unwilling to share her White House spotlight with Grace. Such a pity. It would have been a historic social tableau, spiced by the inevitable comparison of the world's two ruling beauties and their debonair spouses—Her Serene Highness, whose title was for life, dependent upon no scruffy precincts in Chicago, and Jacqueline, the American First Lady who had conquered the capitals of Europe. The White House can handle two princesses with glamour to spare; Jackie should have bent over backward to do it.

During the Truman administration there was another scintillating woman who was not allowed to enter the White House: Clare Boothe Luce, playwright, Republican congresswoman, and wife of the powerful founder of the Time-Life publishing empire, Henry Luce. Comely Clare, whose tongue was as swift and lethal as a viper's, had scorned Bess Truman as "the ersatz First Lady." Harry, ever the gallant husband, told Luce, "As long as I am in residence here, she'll not be a guest in the White House." But after seven years in the cold, Clare was back, a friend of Eisenhower, who then named her Ambassador to Italy.

The deliberate insult of being banned from White House social events can rise to the level of a political crisis. Mary Todd Lincoln, always more emotional than political, learned her limits when she struck Senator and Mrs. William Sprague from an official dinner list. The reason: she loathed Kate Chase Sprague, daughter of powerful Treasury Secretary Salmon P. Chase. But an official dinner is a measure of power, and on that occasion the President, already at odds with much of his cabinet, overruled his petulant Mary.

From the time L'Enfant sketched his first tentative plan for the President's House, entertaining was to be an important function, and it was suitable that the mansion's initial social event would be a reception for members of Congress. In the Christmas season of 1800 Abigail Adams, wearing her finest brocade gown and best face, made them welcome. Twenty cords of wood were burned in a hopeless effort to warm the second-floor reception room (the first floor was not yet usable), which she had brightened with crimson draperies and upholstery, and she had rounded up musicians for piano, harp, and guitar. Even with her best efforts, it was a frosty party—politically as well as socially.

Though she was miserable in the unfriendly, unfinished President's House and heartbroken by her husband's defeat, Abigail summed up the importance of entertaining in a comment that could apply to the White House today: "More is to be performed by way of negotiations many times at one of these entertainments than at twenty serious conversations."

Abigail and John followed the stiffly correct style of the Washingtons, which had been patterned after the courts of Europe. The presidential couple set a regal tone: standing on a raised platform, Adams greeted each guest with a courtly bow, just the way President Washington had done, and Abigail, just like Martha, received seated at his side. Then came Jefferson and a reversal of style; he preferred frequent small dinners at a round table, to encourage good conversation and minimize rank. The wines were French, and in the hands of his French chef the menu reflected the President's interest in new foods from various countries—macaroni from Italy, waffles from Holland, and the recipe for a new French concoction called ice cream. When the elegant widower discontinued Abigail's weekly "levees," the ladies of Washington protested—in person. He charmed them but did not back down.

Any mention of White House entertaining leads directly to Dolley Madison, the first true chatelaine of the White House, the style setter, the political wife, the magnet. With the eight Jefferson years as her apprenticeship, Dolley took over with confidence in 1809, making the President's House the hub of the still-unformed capital city. Along with many formal dinners, each Wednesday evening she held open house, called "drawing rooms," and no matter that they were hot and overcrowded, no one dared miss them because *everybody* was there. In Dolley's drawing room political adversaries conversed amiably and gossip flowed like her well-laced punch; the ladies, decked out in the latest fashion (Dolley liked her turbans twisted very high and her Paris gowns cut very low), mingled freely and joined discussions of the day's news. Her dinners were crowned with ice cream in various flavors and colors, which was a great success. She would be pleased that two hundred years later a popular brand of ice cream is named for her, even though the label misspells Dolley.

This first true First Lady was one of those charmers who know they are naturally irresistible and thus become even more irresistible. Margaret Bayard Smith, the voice of Washington's early social scene, extolled her "unassuming dignity, sweetness, grace. It seems to me that such manners would disarm envy itself, and conciliate even enemies." Dolley turned her social graces to those purposes, always with the primary purpose of bolstering her husband, who could draft the American constitution but could not find words in the drawing room. One guest described him as "a little man with powdered head having an abstracted air . . . but with little flow of courtesy."

Washington Irving, the most celebrated author of the day, wrote of his experience at one of the First Lady's soirées: "I emerged from dirt and darkness into the blazing splendor of Mrs. Madison's drawing room. Here I was most graciously received. . . . Mrs. Madison is a fine, portly, buxom dame who has a smile and a pleasant word for everybody." Her two sisters, vivacious widows who were living in the White House, were "like the two merry wives of Windsor; but as to Jemmy Madison—oh, poor Jemmy!—he's but a withered little apple-john."

In an implicit criticism of Dolley's rather informal style, Elizabeth Monroe, the next First Lady, reverted to Abigail Adams's formality of sixteen years before, and President Monroe even returned to knee breeches. After a dinner with the Monroes, author James Fenimore Cooper wrote, "The whole entertainment might have passed for a better sort of European dinner party at which the guests were too numerous for general or very agreeable discourse." All in all, he found it "rather a cold than a formal affair." The Monroes had missed the point: White House entertaining should be America at its best, not faux European.

Thirty years after Dolley's time, Charles Dickens attended a public levee in the John Tyler White House. (Washington Irving, on his way to his new post as minister to Spain, was again present that evening.) In his fulsome account of the evening, Dickens was impressed that "the company [was] of very many grades and classes. But the decorum and propriety of behaviors were unbroken by any rude or disagreeable incident; and every man . . . appeared to feel that he was a part of the institution and was responsible for its preserving a becoming character. I have seldom respected a public assembly more than I did this eager throng." Dickens had sensed the pride of shared ownership an American feels in visiting the White House, which sets it apart from royal palaces.

Had Dickens attended a reception in John Tyler's final year, 1844, he would have seen quite a change. The President's new young wife, Julia, seated on a raised platform, was gowned in purple velvet with a train and a sparkling headdress that looked suspiciously like a crown, and was surrounded by twelve friends in matching attire, looking suspiciously like ladies-in-waiting. Apparently unaware of the eyebrow-raising effect of this royal tableau, Julia proudly wrote to her mother, "For two hours I remained upon my feet receiving quite in a queenlike state. I have commenced my auspicious reign and am in quiet possession of the Presidential Mansion." It is probably just as well that Julia Tyler was not given a second term.

Dickens might have taken still a different impression of Americans in their White House had he attended the inaugural reception the day James Polk took over from Tyler in 1845. The best-quality hats, cloaks, and canes were filched early in the proceedings, and in the crush of well-wishers a pickpocket lifted the wallet of a commodore, no less. Sadly, it held a letter from former President Jackson, containing a lock of Old Hickory's hair and a note from the legendary Dolley. That evening the Polks—who frowned on dancing—were fêted at two inaugural balls, one at ten dollars a ticket, for the swells, the other at two dollars, for the hoi polloi; the latter degenerated into a free-for-all at the food table.

But that was a tea party compared with Andrew Jackson's inaugural reception, when the White House was overrun by unruly frontiersmen celebrating the first President from the West—Tennessee. In muddy boots they stood on the silk-brocade chairs the better to see Old Hickory, and in general treated the White House like a wayside tavern, draining the barrels of punch intended for toasting the new President and breaking considerable china and glassware in their enthusiasm. The hero-President escaped the melee and repaired to Gadsby's tavern with friends to dine on sirloin of a prize ox.

The arrival of a new President and First Lady always set nineteenth-century Washington atwitter; everyone was eager—even anxious—to learn what kind of family they were, how they would change the White House, and, most particularly, whether their hospitality would be extended to all. Since the President's House was the center of Washington social life, these matters were of great import to the ladies. Never was their concern greater than in 1861, when Abraham and Mary Todd Lincoln, from the West—Illinois— arrived in a city divided by North and South.

Unlike many of her predecessors, Mary, who had encouraged Abe to seek the presidency, was an eager First Lady. Only four days after the inauguration the Lincolns held an open-to-all reception. On that nasty March evening more than a hundred carriages jostled in the driveway, and an hour ahead of the appointed time hundreds more guests thronged the walkways.

At eight o'clock the waiting crowd surged in. The new First Lady from the boondocks was a fashion plate in deep red silk ballooned by hoops, with what would be her signature coronet of fresh flowers in her dark hair. It was the first of the stunning gowns that would later bring much criticism. Soon there was not an inch of open floor space; the crush was such that those trying to leave had to climb out through the large front windows. (At the next large reception that problem was eased by a platform built through a window.)

Not since Jackson's rowdy inaugural reception had there been such a turnout from the western frontier; men in boots and flannel shirts were pressed against Washington matrons in silks and jewels. Trying to retrieve their wraps, guests encountered a repeat of the Polk reception—fine overcoats and hats had been replaced by shabby raincoats and dirty caps. Among the personal notices in the next day's newspapers: "Lost last night at the President's levee, a black ribbed cloth overcoat with velvet collar and facings." Crowds and confusion were evidence that Mary's first party was a success, but with the outbreak of war she would be vilified for entertaining as soldiers died.

AFTER THE WAR, in 1877, the nation was once again at peace, but Washington society was again in an uproar: the word was out that Mrs. Rutherford Hayes was banning all alcohol from the White House, where by custom the multicourse official dinners were washed down with spirits from champagne to brandy. At the Ulysses Grants' dinner welcoming the incoming President and Mrs. Hayes, six wineglasses were clustered at each place—soon to be replaced by a water goblet. The secretary of state, protesting Lucy's decree, threatened to cancel the annual dinner for diplomats unless wine was served, but the First Lady—maliciously dubbed "Lemonade Lucy"—won the showdown.

It was then noticed that the platters of fruit on the Hayes banquet table proved especially popular. "Many wondered why oranges seemed to be altogether preferred," journalist Perley Poore reported with devilish delight. "Concealed within the oranges was delicious frozen punch, a large ingredient of which was strong old Santa Croix rum. Thenceforth (without the knowledge of Mrs. Hayes, of course) Roman punch was served about the middle of the state dinners." The cognoscenti called this pause "the Life-Saving Station." Lemonade Lucy, unaware that she had been tricked by her own serving staff, was pleased that temperance proved so acceptable.

As soon as the Marine Band played "Home, Sweet Home" at 10 P.M. sharp, a few select guests hurried across the way to the secretary of state's office, where he was serving real drinks, without oranges. But later the President shared a secret with his diary: "The joke of the Roman punch was not on us but on the drinking people. My orders were to flavor them rather strongly with the same flavor that is found in Jamaica rum. There was not a drop of spirits in them." In such a farce, who can be sure who got the last laugh? But both sides left the table feeling quite pleased with their clever ploy.

Even with intensive planning and expert service, things do occasionally go awry at White House dinners. At the Kennedys' dinner for President Truman, the grouse resisted all attacks. As Margaret Truman told it, "My knife glanced off mine as if the creature were titanium . . . suddenly another guest's grouse sailed off his plate and onto the floor." At that, Jackie, mortified, had the rest removed. Better for guests to be hungry than humiliated. Had it happened at one of Julia Grant's banquets, the guests would scarcely have noticed one less course among more than a score, but Jackie had pared the courses to a minimal four.

At a Coolidge dinner for the Cuban president, three mishaps occurred in the one meal: one official lost his tie mid-dinner, another broke his front tooth on the food, and yet another broke through the cane seat of his chair, challenging guests to keep a straight face.

And then there was the head of state who had braced himself with several drinks before arriving at the White House for the Kennedys' dinner in his honor. After predinner cocktails, he enthusiastically imbibed the wines at dinner. Along about the salad course he was at least one sheet to the wind. The President, foreseeing disaster, had the butlers speed up the service; after the dessert was bolted in record time, Kennedy steered his guest directly to the ground-floor library. As the President was presenting him with a specially bound copy of his Pulitzer Prize–winning book, *Profiles in Courage,* the honored leader suddenly "dozed off." He was funneled into his waiting limousine, and no one seemed to notice that His Excellency had dropped out of sight. His name remains locked in the White House top-social-secret file.

A social crisis of a different nature erupted at another Kennedy dinner when the visiting head of state wished to bring his girlfriend instead of his wife. Terribly sorry, the social office replied, since only officials of cabinet rank were included in state dinners. The honored guest produced an easy solution: he had the lady friend sworn in as his secretary of state for the night. And so she was there, dressed to the nines and hopefully not making foreign policy between courses. (The LBJs avoided such social potholes by allowing the visitors to pick any ten official guests they wished, be they secretary of state or stately secretary.)

IN THE EARLY days of the White House, when distance made it impossible to summon accomplished performers to Washington, entertainment usually came in the favored home-grown form of dancing. It changed over the years as new styles came along and social views varied, and now and then it was vetoed by the President or his wife.

George Washington, for all his dignity, was an accomplished dancer who called for lively Virginia reels as well as stately minuets. In both dances one could always see light between the partners; it was Dolley Madison who introduced Vienna's romantic waltz, in which the gentleman actually took the lady in his arms. "The waltz," harrumphed a stick-in-the-mud, is "the hugging process set to music." Thirty-some years later, young Julia Tyler added the rollicking polka—which her circumspect successor, Sarah Polk, in turn declared unsuitable for the President's House.

Even as late as 1932 Herbert Hoover, a staunch Quaker, would not enter a room where dancing was in progress, with the single exception of the large party he and his wife gave for their son, Allan, a Harvard graduate student. An orchestra was brought down from New York, and—on that evening only—the President tolerated dancing.

Though Eisenhower did not dance much in the White House, blaming "a bum knee" from early athletics, as a West Point cadet he had been given demerits for dancing the racy—and forbidden— "turkey trot" with a young lady at his class "hop." The medal for the dancingest modern President would go, without question, to Lyndon Johnson, who was always first on the White House dance floor, changing partners often to give as many ladies as possible the cachet of having danced with the President.

Nothing is more fun for White House guests than to be there when the President lets down his hair and enjoys himself. Shortly after his second inauguration in 1973, before the cloud of Watergate settled over the White House, guests saw a different side of Richard Nixon. That evening Nixon, who was usually restrained, even uneasy, at his state dinners, gave what was, in the jargon of 1973, a "swinging party" for royalty of the American kind: the great Duke Ellington.

It was a birthday bash that began with the President awarding the world-famous jazz musician the Medal of Freedom. Then Nixon sat down at the East Room's very grand piano, with its spread-eagle supports, and plunked out "Happy Birthday." The Marine Band—whose talents go far beyond the stirring confines of its founder, John Philip Sousa—jammed, and the Duke sat in for a number or two. The White House rocked. It was a glimpse of the Nixon who might have been, in a second term, freed from his corrosive demons.

In his first hours as President, Bill Clinton showed that his would be a lively White House—he just happened to have his saxophone along and sat in with the bands at his inaugural balls. During his first summer in the Executive Mansion, he staged a jazz festival on the South Lawn, an event first organized by Jimmy Carter. Clinton, of course, whipped out his sax again to play with such jazz greats as Wynton Marsalis, Thelonious Monk Jr., and Illinois Jacquet and schmoozed with all. He hit the notes and talked the talk—commenting, quite knowledgeably, "Tunes in three-quarter time have a built-in rhythmic excitement—you just have to ride them." The impressive roster of performers, reveling in the White House gig, were awed that the President could not only actually play, he could improvise.

Clinton genuinely enjoyed contemporary musicians of every beat. Among the string of headliners during his tenure was Willie Nelson, wearing his signature undershirt and headband, who settled in for a night in the Lincoln Bedroom. The rough-hewn country singer and protester had been around for a long time and had seen a lot of America, but after that night he was starry-eyed: "I felt like I was in a sacred place. When you are there, you feel elevated."

Within the fairly rigid social schedule and natural formality of the White House, every President and First Lady have put their personal imprint on entertaining. For both sets of Roosevelts, born into New York's top drawer, the White House was an extension of the life they had always known. Especially in their early White House years, Eleanor entertained constantly, her guests ranging from haute society to earnest social activists, while ending the huge and gener-ally meaningless public receptions. (Her predecessor, Lou Hoover, had added to that problem by sending engraved invitations, hand-delivered, to every person who left a card, a gracious and costly ges-ture that led to lines curling around the block and a sore hand for the First Lady.)

In the war years Eleanor introduced a different kind of entertaining—three days a week young servicemen from nearby bases flocked to the South Lawn. They chatted and laughed with the First Lady and wolfed down pretzels, chips, cookies, punch, and beer, enjoying a welcome break from the mess hall. They were a dif-ferent kind of White House guest, and the staff loved helping "our boys" have a good time. But their tolerance for the First Lady's egal-itarianism was stretched to the limit when she invited girls from a Washington reformatory to a garden party. Despite the staff's dis-dain, the young inmates behaved with propriety—for one Cinderella afternoon they were welcomed as young ladies by the most impor-tant woman in America.

At the White House state dinners have always been a major produc-tion, but the present President Bush has put his strong stamp on state entertaining: he likes it ranch-style. In his first term he all but aban-doned White House state dinners—holding only four compared with twenty in his father's four years—in favor of visits to his Texas ranch. There, in jeans and cowboy boots, he greets his VIP guests at the heli-copter pad in his white pickup truck and gives them a taste of Texas. For the first state visitors at the ranch, Russian President Vladimir Putin and his wife, dinner was served from a chuck wagon on the lawn, with "real" cowboys doing the cooking—mesquite-smoked pep-pered beef tenderloin and smoked catfish. Entertainment was pro-vided by what Laura called "a great little Texas swing band who sing those great western songs like 'Drifting Along with the Tumbling Tumbleweeds.'" To finish off the evening, the Putins were taught a country-and-western dance, the "cotton-eyed Joe."

The Bushes' first female state visitor, however, was given the full treatment: Philippine President Gloria Macapagal Arroyo was welcomed on the South Lawn with bands playing and flags flying; she was toasted at a black-tie dinner featured Maine scallops, Maryland crab, and lamb. Still, there was a touch of Texas—the entertainer was a soprano from Midland, Texas, Bush's early Texas hometown. Crawford offers a genuine slice of Texas, but nothing matches the White House as the power backdrop beamed by satellite to the visitor's constituents back home.

Even such seemingly insignificant decisions as the clothes he wears reflect a President's attitude toward the White House. Eisenhower brought with him a general's regard for the trappings of authority; he rejected the trend toward informality, hewing to the vanishing tradition of white tie and tails, even though he felt that white tie made for "a stuffy evening." Kennedy, a new generation, relaxed the official dress code to black tie. (The Bush twins would be stunned to learn that in the mid-1920s President Coolidge ordered his college-age son to wear a tuxedo to dinner where there would be no guests, just his parents, because "this is the President's House.")

When the Soviet Union's rumbustious Nikita Khrushchev paid a historic visit in 1959, the depth of the cold war, he refused to wear white tie or black tie, the ultimate mark of the bourgeoisie, and came to dinner in a business suit; his plain wife, Nina, wore a street-length dress, relieving its proletariat dreariness with a diamond-and-emerald brooch. Eisenhower, refusing to defer to Khrushchev, was every inch the capitalist in white tie, tails, and decorations, and Mamie, making her own political statement, chose a regal gown of gold brocade, with dazzling diamond earrings and necklace. Washington got her message and applauded it.

A footnote to that visit: some three decades later Khrushchev's son, Sergei, became an American citizen and was welcomed by Eisenhower's granddaughter, Susan, President of the Eisenhower Institute, a Washington think tank. Susan is an expert on Russia's nuclear nonproliferation programs, her husband, Russian physicist Roald Sagdeev, who served in the Supreme Soviet, works with her in the institute. Susan proudly notes that he was the first person in the history of the Soviet Union to vote against a piece of legislation. Her mission, rooted in her grandfather's legacy in foreign policy, is unquestionably influenced by her eight years as a White House grandchild.

Both Truman, with his love of history, and Eisenhower, with his lifetime of military service, were guardians of tradition. Bracketed by two famous names from great wealth, the two plain-folks Presidents who had grown up as small-town boys in struggling homes—the Eisenhowers mostly strapped for cash, the Trumans knowing good times and bad—demonstrated particular concern about the social standards of the White House. "Truman," Chief Usher West noted with approval, "brought back all the pageantry, all the formality, all the pomp that we had all but forgotten how to execute." Though Eisenhower differed with Truman on policies and disliked him personally, he stood firm with the erstwhile major on pomp and ceremony in the historic house they both revered.

BEFORE THE INAUGURATION date was moved from March 4 to January 20 in Franklin Roosevelt's tenure, a defeated president and his wife were obliged to go through the White House social season as lame ducks, an awkward circumstance that did not make for lively parties. John Quincy Adams (who had declared the increasingly crowded levees "insupportable to me") and his wife were determined to leave with head held high following his defeat at the hands of his hated enemy, Andrew Jackson. At their final White House reception, Washington's shrewd Margaret Bayard Smith peered behind their mask: "[They] came out in a brilliant masquerade dress of social, gay, frank, cordial manners. What a change from the silent, repulsive, haughty reserve by which they have hitherto been distinguished." There was dancing and music, but in Mrs. Smith's view,

"Mr. and Mrs. Adams have gone a little too far in this *assumed* gaiety."

In 1869 Martha Johnson Patterson, Andrew Johnson's daughter and much praised White House hostess, was showing the elaborate preparations for their final White House dinner to reporter Laura Carter Holloway, an old friend. "I am glad this is the last of entertainments," she confessed. "It suits me better to be quiet and in my own home. I am indifferent to them, so it is well it is almost over." Her relief at returning to a private life has been shared by many First Ladies—the demands of one of the world's most glamorous settings can weary even the most dedicated hostess.

An exiting First Lady will be going back to dinners without touchy protocol, to informal parties with guests of her own choosing and no lines of hard-gripping hands to shake (Lincoln called hand shaking "harder work than rail splitting"), to conversation free from fear of blunders, to no more carping in the press, no need for 4,100 pieces of china, no need to count the silver. All in all—though she may have to cope with after-party cleanup—she will welcome the return to the freedom of private life.

And yet . . . now and then, after the coachmen have turned back into mice, as Lady Bird put it, she will miss the elegant dinners, the perfect service, the beautiful clothes, the history-making guests and unforgettable performances. She can feel proud, quite rightly, that she gave thousands of visitors from the world over a warm memory of the American President's House.

ALONG WITH THE harassed White House staff, a First Lady will remember other events, other guests she might like to forget. President George W. Bush, presiding over a small luncheon briefing with television anchors, turned his tweakish humor on ABC anchor George Stephanopoulos, a former Clinton aide: "Welcome back to the White House, George. We'll have to make sure we count the silverware." He was teasing, of course, but in fact the White House

staff does indeed count the silverware—not just after a Stephanopoulos visit, one assumes, but after virtually every event.

Some White House guests apparently feel that anything that can
be tucked into purse or pocket is fair game as a souvenir. Luci Baines
Johnson shakes her head in disbelief when asked about pilfering:
"You wouldn't believe how many spoons disappear! We put out
matchbooks stamped 'The White House,' hoping they would serve
the purpose." Matchbooks may be taken with impunity. At one of
his first-term dinners President Bush handed one to a young lady at
his table—"Here, honey, take this—it'll be a great souvenir for your
grandchildren." Even greater when she tells them who gave it to her.

This theft in the name of souvenirs is not a sign of declining
modern standards. In Lincoln's day guests were less secretive, more
rapacious—they stripped pieces of the velvet coverings from the
walls, ripped fragments of lace from the curtains, snatched tassels
from draperies, and even removed brackets from windows. "How
can they do that?" Lincoln, dismayed, said to a friend.

Head Butler Fields recalled an afternoon tea in the FDR days
when "we lost a gross of linen napkins" and another event when six
dozen teaspoons walked away. The easiest prey were the tiny salt
spoons used at formal dinners, which, said a housekeeper, "just
melted away." (Occasionally, small articles came back in the mail,
anonymously, suggesting an attack of conscience.) After so many
engraved silver pieces had to be replaced over and over, less expen-
sive flatware was substituted—and if some perfectionists sniff that
the quality is not up to White House standards, they should know
that neither are some of the guests.

Silver spoons were small potatoes for bolder souvenir hunters of
the past. At one FDR function guests somehow made away with
two large silver trays and a sizable silver bowl, one of a special White
House set of four. Eleanor Roosevelt could only marvel at how such
bulky pieces could be hidden under a coat.

But the gold medal for the most brazen "souvenir" episode should
go to the houseguest who stuffed a fourteen-inch silver tray

engraved "The President's House, 1898" into her suitcase—and then asked the maid to finish packing for her. The maid, shocked and bewildered, turned to Butler Fields, who solved the problem: "Remove the tray quietly and say nothing." He knew the guest would not, could not, mention it. Though you can't call the cops when the lady in your White House guest room is, bluntly, a thief, you do what you can to guard the property. One hopes she was not asked back. In these days of metal detectors, perhaps White House guests should be required to pass through one on their way out.

While party guests have filched silver spoons and spirited away silver trays, in the years when anyone was free to roam the White House grounds, some took the liberty of digging up plants to enhance their own gardens. But most imaginative were strollers during Zachary Taylor's brief presidency who boldly plucked a hair from the tail of Old Whitey, the beloved steed that had safely carried General Taylor into battle against Mexico's troops. The President rewarded him with the freedom to graze on the White House lawns and an occasional outing for just the two of them—Old Rough and Ready and Old Whitey reliving old times together. In 1850 President Taylor's funeral caisson was pulled by eight fine white horses, but immediately following the bier, in a place of honor, was one more horse, Old Whitey.

GUESTS WHO HAVE contributed most to White House lore are those who made themselves most at home, disdaining the old rule that "guests, like fish, begin to smell after three days." More than sixty years have passed and there's been no equal to the tales surrounding the secret White House visits of Winston Churchill, Britain's great wartime leader, staunchest ally and close friend of Franklin Roosevelt.

First, it must be understood that whatever the whim or habit, a White House guest must be made to feel at home. Churchill's habit of wandering about unencumbered by clothing occasionally pushed that edict to the limit among the Executive Mansion staff, but in the best tradition of domestic service, they looked at their shoes and tried not to notice.

On his first visit Churchill tested beds like Goldilocks: the Lincoln bed was too hard; the queen's bed was just right. Another room was taken over for a map room where he could track his British troops. At breakfast, scotch, or perhaps sherry, replaced orange juice; lunch called for more scotch; champagne went well with dinner, followed by brandy, then more scotch. The staff marveled that he never appeared to be intoxicated.

Churchill's White House visits lasted as long as twenty-four days; his entourage of twenty to forty aides overflowed the usual bedrooms. He could have been a burden, but the Roosevelts enjoyed his lively conversation and wit, even his proclivity for staying up until all hours, talking over numerous brandies. While Churchill refreshed himself with afternoon naps, the President had to deal with the unending flow of the country's domestic issues. After those visits, Eleanor noted, it took her husband several days to catch up on his sleep.

In those wartime years so much had to be kept secret that the White House staff could only guess at the identity of the "Mr. Brown" ensconced in a choice suite. Then Butler Fields saw the face: it was Stalin's right-hand man, Foreign Minister V. M. Molotov. The White House valet, unpacking "Mr. Brown's" bags, reported to Chief Usher West in fright, "He's got a gun in his suitcase! What shall I do?" The Secret Service settled the problem: "Just don't mention it." (At least it was not the lethal weapon named in his honor—the Molotov cocktail.)

One day Fields, told to summon "Mr. Brown's" valet, knocked on the assigned door. As he recounted the moment, "For a second I was speechless, for standing before me was a nude woman with a bath towel wrapped—of all places—around her head. . . . She said, 'Ya?' I said, 'Mr. Molotov wants his valet.' Again she said 'Ya.' I turned my back . . . and shortly she had a robe on, still with the towel around her head, and I led her to Mr. Molotov's room. I would like to think

I acted just as nonchalant as she did." The White House requires considerable aplomb in expecting the unexpected.

IN THE FDR years the White House became quite blasé about the tide of VIP guests—presidents and prime ministers and assorted royalty. But when the guests were Their Majesties, King George VI and Queen Elizabeth of Great Britain and those far-flung corners of empire that still answered to the Crown in 1939, the White House was agog. Weeks in advance the domestic staff were aflutter with preparations for the Ultimate Guests. The two choicest White House suites were done over completely. In all guest rooms new curtains were hung, new towels, blankets, and shower curtains were bought, and whatever looked the least bit tatty was replaced. To avoid any potential problem, the White House obtained samples of London drinking water, which was then analyzed by government chemists—and, lo, White House water was reconstituted to be fit for a king.

The general anxiety was heightened by several pages of instructions from Buckingham Palace, detailing the requirements of the royal couple, their "suite" of titled attendants, a battery of personal servants, and the servants' servants. (Even the bootblack arrived with an assistant.) The White House, modest in size compared with the palaces of Europe, was stretched to its limits. The room-by-room list was exhaustive, both for principals and for servants: "an inkstand with blue ink, thin penholder . . . red pencil, blue pencil, ordinary black pencils with very sharp points"; in the King's bathroom, four glasses, "one of which is graduated"; silk covers for the blankets and for the Queen an additional silk cover "folded in four on the foot of the bed, with one corner turned up"; a half-liter thermos of hot milk nightly for Lady-in-Waiting Countess Spencer (Princess Diana's grandmother); muslin cloths to cover garments hung in closets; ham sandwiches for a midnight snack . . . the list went on and on, right down to hot-water bottles for pre-air-conditioned Washington on June days topping ninety-five degrees.

A number of needs were more substantial: a triple-mirror dressing table for His Majesty, "high enough to enable contemplating oneself when standing"; for Her Majesty (who would become the much loved Queen Mum) "a dressing table perfectly lighted day and night, with armchair of corresponding height"; a special room where a maid could press the Queen's gowns "without the attendance of any outsider," a similar room for the King's valet, to be equipped with "a large solid table for cleaning shoes"; and "very comfortable settees" everywhere. Small wonder that the White House was flummoxed. One commodity, the instructions noted, would be brought from England: "The King generally brings his own liquor and spirits." But he couldn't tuck in his own fountain pen and very sharp pencil.

The White House staff learned in short order that the trouble with royal guests is the royal servants. The King's valet was vociferous about the quality of food and drink served to him; the personal maid was more imperious than the Queen, ordering her White House counterpart to warm up her lunch for the third time. While the President and First Lady were charmed by the royal couple, the White House staff collectively gritted its teeth. The visiting minions were exercising a trait that is not unusual among underlings appended to power—the tendency to equate themselves with their bosses.

The following generation of British royalty, Queen Elizabeth II and Prince Philip, visiting the Eisenhowers in 1957, were such easy guests that Ike later said if they had stayed a little longer "we would have been calling them Liz and Phil." In 1970 the next royal generation, Prince Charles, then twenty-one, and his sister, Princess Anne, caused a similar hubbub among Washington's young society, with much envy of Tricia Nixon's role as his companion throughout the several days of their visit. The Nixons' dinner for the young royals was a gala event under a garland tent on the South Lawn. Charles was charming, as befits a prince; Anne was prickly, as she would always be.

Many years later, in a private conversation, Charles explained that his sister was unaccustomed to American politicians' way of touching (the royal personage) solicitously, as then Speaker of the House Tip O'Neill did in his friendly big bear way. He also confided that he felt that Tricia Nixon was being "thrown at me." (A year later Tricia married Edward Cox, a *Social Register* New Yorker who is now a corporate attorney, and has lived the very private life she had always wanted; the world knows how Charles's marriage turned out.) One postscript to Charles's visit: in 2004 Julie Nixon Eisenhower revealed a royal secret—in those days Charles traveled with an "old, worn teddy bear."

There was, actually, an even earlier White House visit by British royalty. In 1860 Prince Albert, the playboy Prince of Wales (who would become the playboy King Edward VII), was the first of the four generations of the British royal family who would dazzle the White House. However, he traveled incognito as "Baron Renfrew" (as if that fooled anybody). Albert/Renfrew was accompanied by a retinue of such numbers that President Buchanan had to give up his bedroom and sleep on a cot in a hallway.

Along with formal dinners and an exhausting schedule, the President entertained him—if that is the word—at an open reception to meet "real" Americans. (The invitation was actually printed in the newspapers.) "The rush at the doors was terrible," a New York journalist reported. "Confusion reigned. The Royal party have certainly seen Democracy unshackled for once." (Washington's Perley Poore passed on the rumor that "his lordship had slipped away from his guardian and visited some of the haunts of metropolitan dissipation," a bit of sightseeing the Prince wouldn't have mentioned to his mother, that pillar of probity, Queen Victoria.)

For President Buchanan, whose concerns ran to social rather

than political events, sacrificing his bed was well worth it—as a host, he was brilliant; as a President, he was dismal.

FDR biographer Arthur Schlesinger, Jr., once observed that the President "had an odd weakness for European titles, however seedy." As war engulfed Europe, FDR opened the White House to a number of stranded royalty. Most notable—and not at all seedy—was the sparkling Crown Princess Martha, wife of the heir to the throne of Norway, who was in exile in Britain. At the invitation of the President, Martha took up residence in the White House, along with a countess, a baron, and her three young children, with their nanny. Although there was no evidence of impropriety, the cozy relationship between Princess and President triggered gossip in hothouse Washington, and when Martha finally gave up the White House for an estate in the Maryland suburbs, the two remained close friends with a minimum of speculation.

Throughout the nineteenth century social Washington fell all over itself when meeting a European bearing any title, such as the handsome French "count" who was lionized at all the best parties—until someone discovered that he was the cook at the city's leading hotel. The first monarch to visit the White House was Queen Emma, widow of King Kamehameha IV of the Sandwich Islands (Hawaii), who called on President Andrew Johnson in 1866. She was honored at a formal dinner and an open reception. If guests expected colorful native garb, they were disappointed. The Queen from Exotica wore a black silk gown with a diamond brooch, jet necklace, and white lace veil attached to a jet tiara. In 1874 King David Kalakaua of the Sandwich Islands arrived as the first ruling monarch to visit the White House. Guests at President and Mrs. Grant's state dinner in his honor tried not to gape as the King's aide tasted each of the twenty-some courses before His Majesty took a bite. After that the White House drew the line at tasters: visit the President's House and you take potluck.

As ocean liners grew ever faster and more luxurious and air travel began to shrink the globe, the tide of visiting royalty grew ever greater

and the White House guest book became a lesson in geography—
Russia, Prussia, Greece, Romania, Yugoslavia, Spain (handsome
young bachelor Prince Juan Carlos would later reign in Spain), two
generations of Netherlander queens, the King of Siam (who kept Yul
Brynner onstage for years), Nepal, Ethiopia . . . and more.

In the late 1950s King Saud of Saudi Arabia, who bestowed ex-
pensive watches as freely as candy bars at Halloween, led the new
parade of royal rulers from the Middle East—Iran, Iraq, Jordan—
signaling the growing importance of that volatile region. The stream
of foreign leaders brought more than a frisson of title-dropping; it
was evidence that the White House was increasingly the dominant
world power.

The new visitors introduced the White House to a new and varied set of customs. In the LBJ years, King Faisal, Saud's successor,
arrived with an alimentary ailment that limited him to a menu of
chopped broiled lamb and yogurt, prepared for him by the royal chef
in the Blair House kitchen. The chef then arrived at the White
House kitchen with four black carrying cases, which were placed on
a table directly behind the King, with a royal waiter in attendance.
As each course was served, the waiter removed a plate from a case
and placed it before His Majesty; each plate, apparently, held another dab of lamb and yogurt.

As state visitors came from ever farther afield, a special onus fell
on the White House chef, faced with the restrictions and dictates of
many cultures: no pork for Muslims, no beef for Hindus, kosher
food for Israelis, no meat from web-footed animals for Emperor
Haile Selassie, special diets for many. Some customs introduced even
more delicate problems: How, for example, to deal with the two
wives of the King of Nepal?

In the latter half of the nineteenth century even the most remote
kingdoms were accepting the fact that the upstart American republic
had to be reckoned with. In 1860 the first Japanese diplomats—
two royal princes, a dozen noblemen, and a phalanx of servants—
presented their credentials to President Buchanan. Bystanders at the

White House gates gawked at their magnificent embroidered robes, the two swords at their sides, and their odd little pillbox headgear; the diplomats, in turn, were astonished that the President of the United States mingled with the people. It was the first meeting of ancient East and inscrutable West.

Once the Japanese recognized the United States as a bona fide power, the Chinese were not far behind. Late in the century, the Chinese ambassador in his silken robes caused a stir at White House events: his wife, Madam Wu, literally stopped traffic as she was borne through Washington in her regal sedan chair. Much later, in the mid-twentieth century, newly independent African nations brought yet another wave of exotic diplomats, and a new strain of colorful national dress, to the White House—evidence, once again, that the President's House embraces the changing world.

FROM THE TIME the Trumans moved back across Pennsylvania Avenue into the totally rebuilt White House in early 1952, foreign VIPs have been billeted in Blair House. It is actually a cluster of four fine nineteenth-century brick houses that have been connected to provide a total of 110 rooms; the key mansion was built in 1824 by Francis Blair, a prominent editor from an old Maryland family with a shrewd feel for both politics and property. In a city where proximity to power is second only to power itself, he selected a prime location and made the most of it.

Andrew Jackson used to stride across the street in the evenings to talk politics with Blair and a few intimates; on cold winter nights they drew their chairs around the kitchen fireplace—the origin of the term "kitchen cabinet" in the political vocabulary. Martin Van Buren, who followed Old Hickory into the White House, continued to consult Blair almost daily, and it was there that Blair, acting for President Lincoln on April 18, 1861, offered the command of the Union Army to General Robert E. Lee, admired as a brilliant officer of unblemished character. Lee declined—he could not fight against his native Virginia, where generations of his aristocratic family had been intertwined with its history since 1642. Lee's decision could have affected history; his future in the nation's leadership would surely have been a different story had he exercised his prowess as a general on the Union side.

It was while the Trumans were living in Blair House in 1950 that the President narrowly escaped assassination at the hands of two Puerto Rican nationalists. Attempting to shoot their way into the house, one fired point-blank at White House policeman Leslie Coffelt; heroically, the dying Coffelt was able to draw his pistol and kill the attacker with a single bullet to the head. It was a chillingly close call for the President.

Used primarily for foreign VIPs, Blair House comes under the jurisdiction of the chief of protocol, a post held in the Reagan years by Selwa—always known as "Lucky"—Roosevelt, a White House "in-law" as the wife of Theodore Roosevelt's grandson, Archibald junior. She tells of the guest mansion's special aura: "An army of ghosts inhabit those rooms. Sometimes when working there late at night, I could sense their presence."

Among her unforgettable, if less inspiring, Blair House memories were the monarch who brought his own vibrating bed and had his guards sleep on eighteen mattresses on the floor, five of them dozing naked outside the Blair House manager's door; other royals installed their traveling thrones and exercise equipment; and still other visitors came with their own chefs and truckloads of food and equipment. (One smuggled in meat full of maggots, which Department of Agriculture inspectors confiscated, even burning the suitcases that transported it.) The Blair House staff has perfected a mien of no surprise, but it was not easy when a Middle Eastern potentate's aides were cooking on a brazier amid the exquisite antiques. A frequent and particularly welcome Blair House guest was the late King Hussein of Jordan, a great friend of successive administrations.

The kitchen staff was always ready for his request—a good American hamburger.

THIS STEADY FLOW of visitors might suggest that anyone with a White House connection yearns for the chance to visit. That's not the case; some of the closest family members are reluctant guests. In 1977 President Carter's mother, the candid, breezy Miss Lillian, came up from Georgia now and then, without much enthusiasm. Bluntly declaring the White House "boring," she complained, "It's like a museum. I hate it there. I can't do anything, can't see anything. I don't get to meet anyone. I don't even see Jimmy that often." He wisely solved the problem by sending her—nearly eighty years old at the time—on an extensive goodwill mission overseas, which she loved and carried out with panache, even swapping a little joke with the Pope.

At a private audience with Pope Paul VI, the solemn pontiff, noting that he was eighty-one, stated, "I am now ready to meet God." The ebullient Miss Lillian replied, "When you see God, please mention my name." The amused Pope countered, "And you tell Him about me." More seriously, she asked him to pray for rain in drought-ravaged Africa, next on her tour. In each country she visited she was met by deluges. "It rained so much," she said, "I thought I better call the Pope and tell him to lay off." One other prayer she asked of him: "For human rights in my own country, the Deep South."

Miss Lillian would have been a kindred spirit to President Fillmore's father, who paid a short visit to the White House soon after his son became President upon the death of Zachary Taylor. Asked to stay longer, the eighty-year-old "Squire Fillmore" was not tempted: "No, I will go. I don't like it here; it isn't a good place to live; it isn't a good place for Millard—I wish he was at home in Buffalo." His son, the second of nine Vice Presidents who have become accidental Presidents, was back in Buffalo all too soon; though he yearned for four more years in the White House, his party refused to nominate him.

Ronald Reagan's older brother, Neil—always known as "Moon"—refused to stay in the White House during the first Reagan inaugural, preferring to be put up at Blair House. The two brothers had always had an edgy relationship, which grew more tenuous as Neil, a successful advertising executive, and Nancy Reagan failed to mesh as in-laws. With Blair House under renovation at the second inauguration, Neil insisted that he and his wife stay at a hotel. Heavy persuasion by Reagan aides finally convinced him that refusing to stay in the President's House would make a negative family story, so the reluctant Neil Reagans stayed in the White House for the shortest acceptable visit.

Then there was Sam Houston Johnson, Lyndon Johnson's wayward younger brother. In an effort to forestall another of his scrapes, LBJ virtually ordered him to move into a third-floor suite in the White House. It was a kind of elegant house arrest, with the Secret Service keeping an eye on him beyond the White House gates. To Sam Houston the White House, for all its emoluments, was a prison; returning to the mansion at night, he would pretend to the guards that he was being brought back in handcuffs. It was not altogether a joke.

MOST WHITE HOUSE VIP guests have been pleasant and fairly considerate. Not Madame Chiang Kai-shek, wife of China's generalissimo, one of the four Allies in World War II. In 1943, during FDR's third term, when the war was still an uncertain struggle, "the Dragon Lady," as she was called behind her back, looked upon the White House as her personal guest quarters. She demanded for herself all the deference due her husband and thought nothing of checking in for lengthy visits with aides, a personal maid, two nurses, niece, nephew, and bodyguard, with others in her entourage of forty put up at the Chinese embassy.

Born to wealth and privilege in China (and a graduate of Welles-ley College), Madame was more persnickety than the legendary pea-averse princess. She brought her own silk sheets and lace-edged blanket cover, which were to be hand-laundered daily and changed "from top to bottom," Head Butler Fields noted tartly, "even if she went to bed for only ten or fifteen minutes, four and five times a day." The mansion housekeeper snapped, "It was very fussy and ex-pensive, her getting in and out—she wasn't democratic like the Queen." Madame's staff imperiously clapped hands to summon White House servants to meet their mistress's many demands. "They think they're in China calling the coolies," fumed a maid. Chief Usher West noticed that "the White House maids were never so happy to see anybody leave."

More significant, Winston Churchill moved out of the White House when Madame Chiang was coming, to avoid having to speak to her. Fields never forgot Churchill growling, "I know I couldn't stand that woman, and I might not be able to hold my temper—or rather my tongue."

When she died in 2003 at age 106, in her Manhattan apartment, Madame Chiang's obituary described her as "imperious, hard-boiled and calculating." Sixty years earlier the White House staff would have said the same thing. Enduring those demanding visits, even Eleanor Roosevelt's forbearance ran dry—in her memoirs she ob-served that the American-educated Madame Chiang "could talk very convincingly about democracy . . . but hasn't any idea how to live it."

ten

Intramural Skirmishes

————⇒◦⇐————

I T IS A volatile matchup: the White House versus the families who live in it. Now and then disagreements that would go unnoticed in any other setting blaze into intramural warfare. Custom clashes against change; personalities lock horns in showdowns; a vixen brings down a cabinet.

Trouble was brewing in Andrew Jackson's White House. Old Hickory had bested adversaries on the dueling grounds, quelled the Seminoles in Florida, and, outnumbered two to one, vanquished the British at New Orleans. But he met his match in a phalanx of White House females arrayed against him over a matter of love. The standoff destroyed the Jackson cabinet, in essence bringing down the executive branch of government.

The fracas erupted when Jackson's friend and campaign manager, General John Eaton, married Peggy Timberlake, a young widow whose father was landlord of Washington's most popular inn. She sometimes greeted customers in the tavern, a role her enemies—and they were many—twisted into "barmaid." John and Peg's love story was complicated by the fact that she was married to John Timberlake, a naval officer, until he was found dead in a Mediterranean port—a suicide, whispered Washington gossips, committed after hearing that his wife was having an affair with Eaton. According to social arbiter Margaret Bayard Smith, "Public opinion will not allow General Eaton holding a place which would bring his wife into society." Most emphatically not at the White House.

Public opinion might not allow it, but President Jackson did. Without blinking, he named Eaton secretary of war, over strong objections from Vice President John Calhoun—and from Mrs. Calhoun, who flounced off to her South Carolina home rather than mix with that woman in the White House circle. If Peggy was present, cabinet wives walked out of the room or cut her dead. When two leading ministers attacked from the pulpit, the feisty Jackson counterattacked, demanding their evidence. Defiantly, he invited Peg even more often to White House dinners, while other cabinet wives seethed. "Enchanting, ambitious and unscrupulous," observed Perley Poore, "she soon held the old hero completely under her influence."

Moved by emotions stronger than logic, the President must have seen the attacks on Peggy Eaton as a repeat of the vicious slanders that in his mind had killed his Rachel. "The gallant old soldier," Poore reported, "swore 'by the Eternal' that the scandalmongers would not triumph over his 'little friend Peg.'" As the White House social issue escalated into a crippling crisis, Jackson used resignations as a solution—cabinet members who did not resign were forced out. A totally new cabinet was formed, the first and only time an American President dumped his entire cabinet—friends and foes alike—in his first term.

Peggy Eaton's social role in the White House triggered a cataclysm across the Jackson administration and the national government. How did she do it? Poore, abandoning his reporter's objectivity, was besotted by her charms: "Her form, of medium height, straight and delicate, was of perfect proportions. Her skin was delicate white, tinged with red. . . . Her perfect nose and finely curved mouth, with a firm, round chin, completed a profile of faultless outlines." Every day the beguiling Mrs. Eaton visited the President in the White

House, sharing with her champion "a fresh story of the insults paid her," and the echo of Rachel's woes further endeared Peg to him.

After being "decontaminated" in lesser positions, John Eaton was named minister to Spain, where their daughter elevated Peg to greater social status by marrying a duke. After Eaton's death, Peg, by then sixty-one, made a major misstep in marrying a dance teacher and deeding over to him all of her considerable holdings. Within months he eloped to Italy with a much younger woman—Peg's own granddaughter. The impecunious but irresistibly infamous Peggy Eaton spent her last years in Washington, where by then the scandal that rocked the government had faded into a spicy chapter of history. It would have pleased Peg—and Old Hickory, too—that at her funeral a spray of flowers was sent by President and Mrs. Hayes, from the White House she had torn asunder.

AT LEAST ONE intramural dispute had an international dimension. The first British minister to the United States, Anthony Merry, arrived at the White House to present his credentials to President Jefferson decked out in full diplomatic finery—plumed hat, splendid sword, indigo-blue coat festooned with gold braid. He then fixed on the President, who, Merry told his friends, "was not merely in undress, but actually standing in slippers down at the heels, and both pantaloons, coat, and under-clothes indicative of utter slovenliness and indifference to appearances and in a state of negligence actually studied." Even worse, the shocked ambassador reported, throughout their meeting the President flipped one of his slippers in the air and caught it with his big toe.

Thomas Jefferson, a cultivated Virginian and Francophile who had spent years in Paris and knew the fine points of fashion, held little affection for the English. Never would he forget that France came to the aid of America when the outcome of the Revolution was in peril; nor had he forgotten that in London in 1786 King

George III had ostentatiously turned his back on him. The President's cavalier attitude and disheveled dress were obviously an intended rebuff.

Three days later tongues were wagging over the President's new breach of protocol. The requisite White House dinner in honor of the new British diplomat had turned into a battlefield: at a time when Britain and France were at war, Jefferson, incredibly, had also invited the ranking French diplomat. Even more shocking, the President ignored Mrs. Merry and instead escorted Dolley Madison, whose husband was then secretary of state, into the dining room. At this studied insult, the Spanish minister's wife whispered to her friend Dolley, "This will be the cause of war!"

The Merrys, seething, stalked out immediately after dinner. In his report to London, the envoy sneered, "The excess of democratic ferment in this people is conspicuously evinced by the dregs having gone to the top." When Secretary of State Madison repeated the same insulting pattern at his own dinner, Ambassador Merry forced a cabinet wife to relinquish her place to Mrs. Merry. She in turn spread it about that the evening was "more like a harvest home supper than the entertainment of a Secretary of State," to which Dolley coolly replied, "The profusion of my table . . . arises from the happy circumstance of abundance and prosperity in our country." It was war, for certain. "Washington society," the French minister reported to Paris with undisguised glee, "is turned upside down."

The social battle had substantive effects as Merry's enmity toward the President led him to consistently misinterpret American issues: he advised the King that the new republic would fall apart, urged Britain to thwart the Louisiana Purchase, and, in the view of some historians, later prolonged the War of 1812. The battle of the White House dinner table extended far beyond dessert.

The Merrys soon exiled themselves to Philadelphia, away from the snapping turtles of the Washington swamps. One suspects that Mr. Jefferson smiled at this development.

"PROTOCOL" . . . "ETIQUETTE." STUFFY matters, prissy words—
they hardly seem of sufficient consequence to touch off a war between the First Family and the capital city, setting cabinet members
and congressmen against the White House, causing a boycott of the
First Lady's social events, enraging embassies, and dragging the secretary of state into the fray. All of which occurred when James and
Elizabeth Monroe moved into the President's House with their two
daughters, Eliza and young Maria.

The ladies of Washington were eagerly anticipating the reopen-
ing of the White House on New Year's Day 1818, more than three
years after its destruction by the British. They were impressed by
Elizabeth Kortright Monroe's regal appearance and European
polish—in Paris she was admired as *"la belle Américaine"*—and expected her to continue Dolley Madison's social whirl. To their disappointment the new First Lady, not in good health, turned over many
of her duties to thirty-one-year-old Eliza, who lived in the Executive Mansion with her husband, George Hay, the President's private
secretary. Eliza was nothing if not self-assured; schooled elegantly
in Paris during her father's diplomatic service in Europe, she claimed
as friends the daughters of the aristocracy, in particular Napoleon's
stepdaughter (who was also his sister-in-law and, in the convolu-
tions of European politics of the time, would briefly be Queen of
Holland). All in all, Eliza felt quite superior to the locals.

The young Mrs. Hay immediately set about overturning White
House social customs, starting with the ritual of social calls, which
had grown to scores a year. Exchanging visits was the primary preoccupation of the ladies of Washington—merely leaving a calling card
with the White House doorman assured at least an invitation to tea
and perhaps even a return visit by the First Lady. Eliza then decreed
that embassy wives must call upon her first, construed as a matter of
deference. That was too much—this upstart of a daughter demanding the same social status as a First Lady. Whereupon, the ambassa-

dors themselves threw down the gauntlet; this went beyond their wives' wrath—national pride was at stake. Eliza, unyielding, retaliated, employing the maximum White House weapon: the President. She urged her father to treat ambassadors with a show of frost.

The ladies of Washington struck back, boycotting the Monroes' weekly informal receptions. The capital's most influential society editor reported with a touch of schadenfreude, "The drawing room of the President was opened last night to a beggarly row of empty chairs. Only five females attended, three of whom were foreigners." White House social events were, de facto, turning into stag affairs, ripping the fabric of Washington entertaining. Those White House "drawing rooms" were more than mere parties; in a capital beginning to mature, they provided the opportunity to meet and mingle—in today's parlance, to network and schmooze—always important in the city where dropping the right name is the coin of the realm.

In that poisonous atmosphere President Monroe assigned the vexing question of protocol to Secretary of State John Quincy Adams, who possessed the fewest social graces of any man around the cabinet table. (He described himself—without contradiction— as "a man of reserved, cold, austere and forbidding manners; my political adversaries say a gloomy misanthrope and my personal enemies, an unsocial savage.") Having drawn the short straw, Adams was forced to cope with "the obstinate little firebrand" Eliza, who kept pestering him about the "paltry passion of precedence"—the keystone of influence—in "this senseless war of etiquette visiting."

The abrasive Eliza detonated yet another social bomb as she planned her charming sister's nuptials, the first White House wedding of a President's daughter. Maria, not quite seventeen, was marrying her first cousin, Samuel Gouverneur, an assistant to her father, in an elegant event replete with four bridesmaids in a state parlor. How could such a happy occasion cause such a storm?

Eliza found a way. The wedding, she declared, would be "New

York style," limited to family and intimate friends, thus insulting the cabinet and diplomats, rudely barring them—and their gifts. (The bride had no say in her wedding plans.) Predictably, they were furious. Louisa Adams, who would be the next First Lady, exploded in her diary: "This woman [Eliza] . . . is so proud and so mean I scarcely ever met such a compound. . . . I never have heard her speak well of any human being." Hardly the qualities of choice for a substitute First Lady.

But by the end of the Monroes' first term the battle had lost its steam. Missing their White House entrée, the ladies once again flocked to the First Lady's entertainments, and at the Monroes' farewell reception in 1825 the crush of guests was large enough to allow General Winfield Scott's pocket to be picked. The Monroes, despite years of experience in the courts of Europe, had to learn an American lesson: the White House is not to be trifled with. Its occupants must never forget that they are only temporary and are well advised to walk softly when they tinker with tradition, no matter how trivial the issue.

To modern ears this brouhaha sounds like the ultimate tempest in the afternoon teacup, but at the time the etiquette war was characterized as "earthquake, upheaval and cyclone." More than a hundred years later, in the Harding and Coolidge era, this social gavotte was still deplored, yet still followed, within limits. Grace Coolidge received her callers in small groups for easy chitchat; the highhanded Florence Harding seated each caller in a small chair directly facing her, allotting five minutes for the guest to try to make conversation, an arrangement akin to a prison visit. Eleanor Roosevelt abolished the practice; she had more useful things to do.

In 1929 Washington was maliciously savoring the Situation. Previously, White House social secretaries had taken refuge in the immutable rules of protocol to settle seating predicaments, but the Hoovers' secretary was floundering in uncharted, shark-infested social waters without protocol as a life preserver.

Before Jackie Kennedy introduced small round tables, the seating chart for an official White House dinner was the map that calibrated precedence and power: Who was below the salt? Who was within buttering-up distance of the President? The State Dining Room itself was a field of honor where jousts were settled according to rules laid down long ago by an official who had likely sought safety in some remote Third World post.

But nothing in the rules covered the clash that had Washington chortling. The adversaries were formidable and implacable: Dolly Gann, Vice President Charles Curtis's official hostess, v. Alice Roosevelt Longworth, wife of the powerful Speaker of the House of Representatives and daughter of the twenty-sixth President, each demanding status second only to the First Lady. The matter had reached an impasse. "Hostesses were afraid to ask Dolly to dinner," said a contemporary socialite. "The State Department refused even to discuss it. The chief of protocol was 'out' whenever anyone called."

Usually Princess Alice, a blithe spirit not given to following rules, would not care a fig about such a boring detail, but it had suddenly become a matter of principle—or pique. She was being upstaged by the Vice President's half-sister—not even a full sister, much less a wife. When he was a senator, the long-widowed Curtis had made his home with Dolly and her husband. Dolly was an Amazon who stood a good head taller and undocumented pounds heavier than her husband. With her penchant for gowns most kindly described as memorable, she had cut a swath through the capital's social scene. But when the Ganns moved with Curtis into the vice presidential apartment in the Mayflower Hotel and Dolly was designated the Vice President's official hostess, the trouble began.

The crux of the standoff was the unwritten, though never unnoted, Washington understanding that no one upstaged Alice Roosevelt Longworth. While the ladies of Washington did not dare cross the razor-tongued Alice, Dolly did, with gusto, enlisting her brother to use his newfound vice presidential clout to press her case. The diplomatic corps, caught between the two forces, acted in unison to side with the Vice President and the formidable Dolly Gann—let the White House handle the crisis as it wished. As an astute observer of the showdown explained, "Protocol is as much the law of social Washington as the Constitution is the law of the land. It settles irrevocably who's who in official life . . . then decides exactly how much more, or less, important everyone is than everyone else." (In the Kennedy years, the French ambassador, with this in mind, was seen changing place cards to upgrade his location at a White House dinner.)

The indomitable Alice, who had never been bested, was forced to retreat. Official wives didn't know whether to cheer or jeer: though they heartily disliked snobbish Alice, they didn't like Dolly, either—and she was usurping a wife's place in the hallowed rules of protocol. So they followed their own interests and sided with Alice. The Hoovers came up with a Solomonesque solution: they added a White House dinner for the Vice President, where Dolly could shine, and a separate dinner for the Speaker of the House, where Princess Alice could reign as queen bee. That settled, the President could get back to the problems of the stock-market crash and the onset of the Great Depression, which did not work out as well.

NOBODY WOULD EVER suggest that Harry Truman was wishy-washy in making up his mind. He was just like an old Missouri mule, some said—when he knew which way he wanted to go, that's the way he would go. Early in 1948 he proposed adding a balcony to the South Portico, the first major alteration to the White House exterior since 1829, when Andrew Jackson added the North Portico, the gracious entrance that became the White House signature. Criticism, predictably, rocketed in from all sides—hostile editorials, ridicule in cartoons, and endless jokes. People resented a President trying to change the White House. Presidents come and go, but the White House is America, permanent.

The *New York Herald Tribune* sniped that Truman had "a lamentable penchant for meddling with an historic structure which the nation prefers as it is." The *Washington Star* sneered that it was fortunate that Truman had not seen the Taj Mahal or the old Moulmein Pagoda, lest he come up with even worse ideas. Others derided it as "Truman's folly." Most significant, the Fine Arts Commission, the White House watchdog, barked a definite "No!"

A resolute Harry was ready for them with a practical argument. The awnings that had to be attached halfway up the columns to shield the state rooms from the blazing south sun "looked always as if they'd caught all the grime and dirt in town," he said, and they cost two thousand dollars every year in upkeep. He calculated that "in eight years the portico will be paid for in awning costs." With his coconspirators in the usher's office, Harry was considerably blunter: "I'm going to preserve the architectural scheme of the White House any damn way I want to!" The Fine Arts Commission, Chief Usher West recalled, "went up in smoke." "The hell with them," Truman said. "I'm going to do it anyway." And he did.

Once in place, the balcony looked as if it had always been there. "And that's the way it ought to be," the President declared with satisfaction. He acknowledged a personal reason for adding the balcony—it would give White House families a private place to enjoy the view and a breeze, a necessity in Washington's suffocating, non-air-conditioned summers. Without it, they were trapped inside. By the time the renovation was completed in 1952, Harry's point about the heat was irrelevant: the new White House was centrally air-conditioned. Still, from that time on, presidential families have enjoyed Harry's balcony, the best seat in town for watching the Fourth of July fireworks.

In retirement, Truman mused that the controversy "didn't surprise me a bit. That was an election year. And people get in an uproar whenever any change is suggested in the White House. My goodness, when Mrs. Fillmore wanted to put in bathtubs . . . people screamed and carried on but eventually everybody calmed down. And I knew the furor over the balcony would calm down, too." He maintained that Jefferson had envisaged a balcony. "And that's where I got the idea, from old Tom Jefferson." Old Tom, he pointed out, used similar balconies in his design for the University of Virginia. Case closed.

Who would deny that the White House is more beautiful today with the curving grace note of Truman's balcony enhancing its South Portico? The President's House is indebted to its thirty-first President's knowledgeable eye, respect for Jefferson, and stubborn will.

"I NEVER COULD understand why there was so much fuss about my balcony," Truman said in retrospect, "not when so many worse things had been done." One of the "worse things" was the plan pushed in 1889 by First Lady Caroline Harrison, wife of Benjamin Harrison, who was bulldozing her way toward her vision of a new and grander mansion on the order of a royal palace. The White House, with ninety years of American history in its bones, was about to be demolished—until Carrie learned a lesson in practical politics. The Speaker of the House, neither history buff nor architecture purist, was the wrecking ball that flattened her grandiose scheme. His veto was frankly political: he was mad at Benjamin Harrison for rejecting his candidate for a cushy federal position, and he got his political revenge through the First Lady. So Caroline settled for renovation, updating the mansion with multiple bathrooms and electricity, which she equated with electrocution.

Only nine years later, the President's House was again endangered when President McKinley proposed a "handsome, permanent memorial" for the centennial celebration in 1900, involving major changes in the White House. Caroline's plans were dusted off, with a few revisions, and a model was placed on display in the East

Room. The redesign was on track until the American Institute of Architects took a close look and declared it a monstrosity. Once again the White House survived an onslaught.

A year into President Eisenhower's second term, plans were afoot to demolish the grand old Executive Office Building, an exuberant nineteenth-century period piece (originally housing the State, War, and Navy Departments) adjacent to the White House, and replace it with a modern office structure. While the President went along with that controversial proposal (which was later scuttled by the Kennedys), he defended the Executive Mansion itself against any threat of change: "We must never minimize what the White House, just as a building, means to America. I have seen strong men come into that building . . . with tears on their cheeks. The White House should never be overshadowed by anything or ill-treated." The twentieth-century general-hero was true to the vision of his eighteenth-century predecessor George Washington. Now safely past its two hundredth year and having weathered repeated assaults, no matter how well intended, the historic mansion seems safe from despoilers.

FDR DEMANDS NEW DEAL
REFUSES SPINACH
CRISIS STRIKES WHITE HOUSE

The newspapers trumpeted it: war had broken out in the White House! The President of the United States was demanding a showdown with the White House commander in chief—the housekeeper. "Liver and beans three days in a row . . . the same salt-fish breakfast four days running" was provocation enough for a counterattack. He mobilized his troops in the press, who ate it up and came back for seconds. His story of war against an internal enemy, Henrietta Nesbitt, made headlines all across America.

When the President declared to the press, "I'm sick of liver and beans!" Mrs. Nesbitt dismissed his outburst as merely a "figure of speech." What's more, she harrumphed in her nanny-knows-best manner, "He was *supposed* to have them!" The way she told it in her thick-headed memoirs, "The President never complained and enjoyed every mouthful. Now, out of a clear sky, he blew up about the food. Whenever he became tense, he would get peevish about his meals." Everyone else in the White House had seen thunderclouds in her clear sky. The President had, in fact, reached his limit of toleration—he could direct World War II, but he was losing his battle against the housekeeper.

"The President would eat all sorts of funny things when he was away from home," Mrs. Nesbitt noted. "Relief, I guess, from the strain of the White House." Perhaps relief from inedible White House food? FDR was anything but a picky eater—he polished off gifts of pigs' feet (even Churchill passed on those), the hindquarters of Greenland musk ox, terrapin, a barrel of stone crabs from Ambassador Joseph Kennedy, five buffalo tongues, and more. That suggests an adventurous appetite that would try anything, so long as it wasn't three straight nights of liver and beans.

The President had some big guns on his side. Novelist Ernest Hemingway, who had dined on army rations during World War I and campfire grub in the heart of Africa, summarized his dinner with the Roosevelts in 1937: "The worst I've ever eaten. We had rainwater soup, followed by rubber squab, a nice wilted salad and a cake some admirer had sent in. An enthusiastic but unskilled admirer."

"The food around here would do justice to the automat," FDR groused. "I wish we could do something about Mrs. Nesbitt, but Mrs. Roosevelt won't hear of it." Eleanor herself cared nothing about food. "She was too much interested in talk to care what she ate—she'd eat anything put before her," Mrs. Nesbitt wrote, in an unwitting insight into her culinary skills.

It seems hard-hearted that the First Lady put Mrs. Nesbitt's complaints above the overworked President's. Perhaps she could not admit the egregious shortcomings of the housekeeper she herself had hired, based solely on their acquaintance at the Hyde Park League of Women Voters and the bread she had baked for the Roosevelts. Mrs. Nesbitt's experience came from cooking for her large family back in Duluth, Minnesota. "I was small town and a homebody," she boasted. "I was noted for my thrift and way with leftovers." The shrewd housekeeper never forgot which side her hard rolls were buttered on—she answered to the First Lady, not the President.

In his third term the President solemnly announced to his secretary, Grace Tully, "I really want to be elected to a fourth term." Grace gasped—this was worldwide news. Then he added, with a triumphant grin, "A fourth term, so I can fire Mrs. Nesbitt!" For ten years the President had skirmished with this dragon of the nether regions and had never won. He could take on Hitler and deal with John L. Lewis, the ferocious head of the coal miners' union, but in a demonstration of the limits of power, Housekeeper Nesbitt had bested the President of the United States.

The President never won that battle. At his death, his sons wasted no time in taking up their late father's cause with the Trumans. On the funeral train returning to Washington from Hyde Park, as Truman himself told the story, with considerable relish, "Mrs. Roosevelt and Franklin Jr. and Elliott came in [the Trumans' compartment] to see us. With their mother sitting right there, they said, 'The first thing we want you to do is fire the housekeeper. She's starved us to death.' And Mrs. Roosevelt said, 'Why, you boys had enough to eat, you know you did.'" Was raising the issue of Mrs. Nesbitt on that day, in that setting, an oblique reproach to their mother for disregarding their exhausted father's comfort?

After the Trumans moved into the White House, Margaret told Mrs. Nesbitt that her father did not like brussels sprouts—whereupon the President was served brussels sprouts for the next three nights. Meanwhile, Mrs. Nesbitt confided to Head Butler Fields that she had had a little talk with the new First Lady: "I just told her that you don't do things that way. I said Mrs. Roosevelt never did things that way." The housekeeper's diplomacy was on the level of her cuisine.

The sticking point came over a stick of rationed butter that the new First Lady had picked up to take to a potluck luncheon. As Chief Usher West related the scene, the imperious Mrs. Nesbitt challenged the First Lady, "Oh, no! We can't let any of our butter go out of the House. We've used up almost all of this month's ration stamps already." With that snappish lecture, the kitchen dictator's goose was cooked. As Truman told it, smiling knowingly, "After a time, Mrs. Nesbitt left. I believe she decided, suddenly, to resign." Quiet Bess Truman had accomplished what President Roosevelt, the most powerful man in the world, could not.

THE CIVIL WAR was four years in the past, and the general who had won the laurels, Ulysses S. Grant, had also won the White House by a hero's landslide. But within the Executive Mansion the North-South conflict continued to flare, pitting the President's father, confirmed Unionist Jesse Grant, against the First Lady's father, unreconstructed Confederate "Colonel" Frederick Dent. In territorial terms Dent, a former slave-holding Missouri rebel, held the high ground: he lived in the White House with his daughter's family, along with the husband who came with the package, the son-in-law he labeled "a turncoat"—the President.

The old faux colonel made himself totally at home, staking out a well-placed chair in the reception room, where politicians and journalists found him a source of colorful anecdotes and insider tidbits. The President's own father, on his frequent visits from Kentucky, would go only so far in accepting the uneasy truce with this symbol of the enemy—Jesse Grant insisted on staying in a hotel rather than sleeping under the White House roof with "that tribe of Dents." In this one battle the general-President hoisted the white flag of truce.

Though the White House does its utmost to cater to the First Family's whims and idiosyncrasies, there are times when it cannot save them from themselves.

eleven

Slings and Arrows

———⊰•⊱———

I F YOU LIVED your life trying to make sure that nobody ever criticized you," Eleanor Roosevelt once commented wryly, "you would probably never get out of bed—and you'd be criticized for that!" In her twelve years in the White House, she became inured to the churlish chorus of disapproval, objections to a First Lady staking out a serious public role of her own and using her position to shine a light into the darker corners of American life. She was ridiculed for going down into a coal mine, scorned for accepting a flower from a child in the slums, lambasted for her fight against lynching and her advocacy for civil rights. But she never wavered.

Sixty years later, long after the women's movement opened new doors to women, not all that much had changed when it came to First Ladies; they were still judged by different standards. Hillary Clinton was asked during a visit to Australia why she seemed to generate such controversy. The only way a President's wife can escape criticism, she replied testily, "is to have a bag over your head when you come out in public, or in some way make it clear you have no opinions and no ideas about anything."

For all its comforts and prestige, life in the White House comes not without cost in ways large and small; it holds the threat of unexpected political potholes, hidden dangers, blazing headlines. The plight of every First Lady is that she is suddenly a national property, to be dissected, analyzed, praised, criticized—and she is helpless to do anything about it.

Even the venerable Martha Washington drew behind-the-glove whispers as she carved out the new role—complaints about the bowing and scraping she required, which smacked of monarchy to her critics.

Eleanor, the social activist, might have secretly enjoyed rattling reactionaries, but for wives and children who have never experienced controversy, public rebukes and ridicule can be devastating. If a First Lady's entertaining is too lavish (Mary Todd Lincoln), there's sniping; if it is a bit frugal (Sarah Polk), there's carping. "Lemonade Lucy" Hayes was faulted for not serving wine; Jackie Kennedy was berated for serving drinks from a bar (one time only—her husband ruled it out).

No matter what she does or doesn't do, a White House occupant can expect disparagement, much of it politically motivated. Superprotective Nancy Reagan became "Lady Macbeth" in a red designer suit. Mamie Eisenhower, as bright-eyed and glowing as a schoolgirl, was dogged by rumors of secret drinking. Abigail Fillmore was always warm and welcoming at White House events—leading the capital's upper crust to sniff, "It is not good form to be so motherly to her guests." Martha Washington was reproved for being "too queenly." Chief of Protocol Lucky Roosevelt came to the conclusion after years of close observation, "Most First Ladies can't win, whatever they do."

Eleanor Roosevelt shrugged off the slings and arrows. "Destructive criticism is always valueless," she declared, "and anyone with common sense soon becomes completely indifferent to it." Her advice to women who want to make a difference: "Develop a skin as thick as a rhinoceros hide." Sound advice, but it doesn't keep the hostility, the nitpicking, and the harsh judgments from hurting.

Hillary Clinton took her role model's "rhinoceros hide" to heart. "[It] had become a mantra for me as I faced one crisis after another," she wrote in her memoirs. "That may have made things endurable, but it didn't make them easy. You don't just wake up one day and say,

'Well, I'm not going to let anything bother me, no matter how vicious or mean-spirited.' It was, for me, an isolating and lonely experience." Hillary acknowledged that she "wasn't ready for how cruel this city can be."

Did Hillary ever consider the fact that she had contributed to the criticism by moving too far out front, too soon? She was a target before she ever crossed the White House portico, from the moment Bill Clinton, as the candidate, declared that the voters would get "two for one"—referring to Hillary, not his running mate, Al Gore. The opposition seized that as a move for a copresidency, a notion that has no moorings in either the Constitution or common sense. Then, after winning the White House, the Clintons made another ill-advised—and unprecedented—move, giving the new First Lady her own office in the West Wing, a choice office usually earmarked for a senior presidential assistant.

"Hillary was in trouble from the minute she decided to have an office in the West Wing," observed a veteran of White House infighting. But she stuck to her guns—and her second-floor office—with her staff parceled out between the East and West Wings.

Putting Hillary in charge of the major task force on health-care reform was yet another red flag for hatemongers who had openly targeted her before the election, and again she played into their hands with some ill-considered decisions. In an effort to counter the criticism, she took the case for health-care reform to the country. In Portland, Oregon, she met cruelty of a different level, threats so violent that on Secret Service advice, she wrote, "for once, I agreed to wear a bulletproof vest. It was one of the few times I felt in real physical danger." As she left, "hundreds of protesters swarmed around the limousine. . . . I underestimated the resistance I would meet as a First Lady with a policy mission."

This postfeminist-generation First Lady met with a cruel, hurtful, even threatening public reaction—and yet, Senator Hillary Rodham Clinton of New York walked away from the White House with the finest consolation prize ever won by a First Lady.

ELEANOR ROOSEVELT HAD set a powerful example for standing up to confrontation. In 1938, when the Daughters of the American Revolution barred world-famous contralto Marian Anderson from performing in their auditorium, Constitution Hall, because she was black, Eleanor could not stand by passively. Defying scurrilous attacks, she arranged for the acclaimed singer to perform at the Lincoln Memorial.

On that Easter Sunday, a crowd of seventy-five thousand filled the vast sweep of the capital's Mall. With the Great Emancipator looking over her shoulder, Marian Anderson's majestic voice rang out in classical works, "America the Beautiful," and Negro spirituals. That day Anderson was more than the "voice heard once in a hundred years," she was the symbol of both the dark undercurrent of the American dream and the promise of change that would come. Eleanor Roosevelt, using the singular power of the White House, made a statement to the nation and the world that needed neither words nor translation.

The following year the First Lady used the White House itself as her spotlight; she invited Anderson to perform, along with white artists, at the state dinner for the King and Queen of England. Even within the Roosevelt White House, there was disapproval of black and white performers appearing on the same program. Eleanor ignored it and formally presented the black artist to Their Majesties. Once again, her message was loud and clear.

But now and then Eleanor lost her White House compass, leaving her open to justified criticism. Archie Roosevelt, Jr., Theodore Roosevelt's grandson and Eleanor's cousin (and later a senior CIA intelligence officer), told of her misstep in "playing godmother" to the American Youth Congress, an organization with Communist-front ties. In 1940, with a group of Harvard students attempting to break the pro-Soviet tilt (the Soviet Union was then an ally of Nazi Germany), young Roosevelt attended an AYC convention in Washington. "I saw Cousin Eleanor sitting in the second row," he wrote, and when his group was thrown out of the meeting, Eleanor sided with the AYC on a technical point.

At a White House reception she hosted for AYC delegates, President Roosevelt denounced the Soviet Union as "a dictatorship as absolute as any in the world" and condemned the AYC's pro-Soviet stance; Eleanor undermined him by commenting, "I don't think you should adopt resolutions on anything you don't believe," tacit encouragement to hold to their pro-Moscow line. More than forty years later, Archie was still dismayed by the continuing "naivete" Cousin Eleanor, the First Lady, had shown regarding the American Youth Congress when it mattered.

It was Eleanor's interest in the AYC that brought Joseph Lash, executive secretary of the American Student Union, into the White House as another of her permanent guests. Chief Usher West observed that young Lash became her closest confidant; she would step across the hall to say good morning, and they would sit in his room talking through the evening hours. The First Lady's relationship with Lash, in his early thirties and unconnected to the White House, touched off gossip that was completely unsubstantiated. (Years later he wrote a prizewinning biography of his mentor.) When Lash was drafted, Eleanor tried unsuccessfully to obtain an officer's commission for him. At first, however, he was stationed at Bolling Field on the edge of Washington, continued to live in the White House, and was chauffeured to his army desk job daily in the First Lady's White House car.

Over the years Eleanor was pelted with criticism—some of it justified—but she never wavered. Chief Usher West understood her: "She was propelled by dedication."

THOUGH PRESIDENTS' WIVES and families are regularly used for image purposes, when it comes to policy making there has always been a dividing line between East and West Wings, a demarcation breached overtly by Eleanor Roosevelt and Hillary Clinton, covertly by Edith Wilson, inadvertently by Rosalyn Carter, and courageously by Lady Bird Johnson.

In 1964, after serving the final year of John F. Kennedy's unfinished term, Lyndon Johnson was running for President on his own. Lady Bird boldly decided to take the blows she knew would come: her husband's historic civil rights legislation had made him an anathema to the South, so she, a Texan with deep southern roots, campaigned through Dixie as his surrogate.

Aboard her own seventeen-car train, the *Lady Bird Special,* she whistle-stopped through eight states for five days, making forty-seven speeches. It was wrenching to witness hard-eyed young people vilifying the gentle First Lady with taunts, jeers, and nasty signs; at one South Carolina train station a young man spit at her. Beautiful Charleston could not have been less hospitable as she toured its storied streets in an open carriage, normally the Chamber of Commerce's dream sequence, to be rebuffed by antebellum mansions with shutters tightly closed and mock real estate signs declaring, THIS HOUSE SOLD ON GOLDWATER.

At a few particularly rough stops she put aside her philosophy that "wisdom lies in knowing when simply to ignore the criticism." She countered the hecklers with an edge in her voice: "You've had your turn. Now it's my turn to say a few words." By the end of the bold gamble, ever-larger crowds turned friendly; nervous politicians put a finger to the wind and jumped aboard. The First Lady's whistle-stop was the biggest and most colorful story of the campaign. Never losing grace and dignity, Lady Bird withstood insults and hatred—and carried most of the South for Lyndon Johnson.

The soft-spoken and issues-minded Rosalynn Carter ran afoul of the policy divide shortly after moving into the White House. Facing a schedule crowded with conferences and speeches related to her longtime activities in the field of mental health, she wanted to be prepared for her new public role: "I told Jimmy I was worried about answering questions I would be asked, and he said, 'Why don't you come to cabinet meetings?'" So she sat inconspicuously with staff members, listened, and learned.

It was as innocuous as that, but snipers took aim at her for over-stepping her role. "There was a lot of criticism," she acknowledges, "but I had long ago decided, do what you think is best." It was, in fact, a good idea: a First Lady should be encouraged to attend as an observer, to be well informed about administration policies, the better to prevent missteps in public situations and to share her husband's concerns. It could have been turned into a positive practice, but perhaps the West Wing staff wasn't ready to accept the concept of a First Lady breaching the east-west boundary.

Surprisingly, there were no complaints back in 1845 when First Lady Sarah Polk played a role of significance to her husband. Having no children to keep her busy as a mother, she became the first working First Lady, serving as her husband's personal secretary. Well educated, she handled correspondence and combed the newspapers for grist for his speeches, which she helped write and edit. It must have been Sarah who encouraged her husband to lend his support to the controversial meeting in Seneca Falls, New York, where in 1848 a few bold women met to organize the long crusade for women's rights; it is highly unlikely that a male assistant would have advised the President to step into that briar patch.

Sarah was known for her intelligent, even clever, responses to gentlemen guests, but not for her own thoughts—that might have been pushing herself a step too far. As further evidence of her tact, Sarah as First Lady happily played second fiddle to eighty-year-old Dolley Madison, still the capital's Premiere Lady, as the star of her White House entertainments.

For Nancy Reagan, the East Wing–West Wing divide was no more than boxes on an organization chart—she had staked out her place on the pillow beside the President. Ronald Reagan was never one to get close to people—Nancy filled his life—and by nature he disliked dealing with personal issues among his staff. Nancy was his shield, his bulwark, his fan club, and she assumed with single-mindedness the role of watchdog, concerned only about total loyalty to her husband.

The West Wing roiled with resentment at the First Lady's thumbs-up, thumbs-down power over her husband's advisers and retaliated with well-planted leaks that were damaging to her. As conventional Nancy stretched the dimensions of a First Lady's role to an unconventional level, they grumbled that the place for the President's wife was in the President's House, not his office.

More than one major player on the Reagan team became past tense due to Nancy, most bluntly Chief of Staff Donald Regan and Secretary of State Alexander Haig. At one rung lower, Helene Von Damm, a presidential assistant who was appointed ambassador to Austria against Nancy's wishes (had there been a West Wing territorial dispute?) wrote that she was forced to resign from her ambassadorial post, not by the President or the secretary of state but by the First Lady. (Van Damm had gotten involved in a tangled romance that made spicy headlines in Vienna.)

Former White House staffers admit that they feared Nancy Reagan, a doe-eyed sliver of titanium. In 2004 she pleaded with Bob Colacello, author of a friendly biography of Ronnie and Nancy, "Please don't make me sound like some kind of master backstage manipulator. Everything I did, I did for Ronnie." But that, of course, was what led to the accusations against her as the power on the pillow.

One of those who ducked behind a pillar to avoid the First Lady was Reagan speechwriter Peggy Noonan. "She was scary to me," Noonan confessed. Twenty years later, long after she had became a friend, the writer asked Nancy, "How did you learn to deal with the criticism?" "You just do, that's all," replied Nancy with resignation. "There's nothing you can do about it. You just hope that people will get to know you and understand you." By then the criticism seemed irrelevant to her life. One thing that both friend and foe accept is that Nancy Reagan's total devotion to Ronnie never waned—it lasted through the long years when he did not recognize her, the Alzheimer's years that robbed her of their golden life together.

From the first days of the Reagan White House, Chief of Protocol Lucky Roosevelt wrote, "It was open season on Nancy Reagan." The potshots came from every direction: aimed at the new $200,000 china service for the White House, though it was a gift, with no public money involved; at the designer clothes she borrowed but did not return or declare as taxable gifts; at the second-rate entertainment at state dinners. (Roosevelt wrote that the President's dislike of classical music meant that "our foreign guests were often subjected to has-been popular singers and other marginal performers who were not up to White House standards.")

Those brouhahas paled beside the revelation that the First Lady had regularly turned to San Francisco astrologer Joan Quigley for advice and precise guidance on White House matters—advice that was unfailingly followed. Interviewed by oral historians Deborah and Gerald Strober, Quigley said, "I did more than fifty or sixty charts for him every year. I was timing the takeoffs and landings of *Air Force One*. I timed all the press conferences." She insisted that she picked the times of the presidential debates.

Lucky Roosevelt had been amazed that Nancy "could overrule the State Department and the NSC [National Security Council] with regard to the dates and the desirability of a [state] visit." Most astonishing were the changes dealing with the signing of a historic bilateral nuclear treaty between the United States and the USSR, delicate arrangements that had been carefully negotiated between American and Soviet officials. The First Lady demanded—and got—a change of date and time: Only later did the American side learn that such a crucial decision was made on the advice of an astrologer. True, Mary Todd Lincoln had held séances with her psychic in the Red Room, but Mary was bordering on lunacy, while Nancy was very much in command.

That fabled command met its test in Raisa Gorbachev. In 1987, in the first faint sign of a thaw in the cold war, the Russians were coming, led by General Secretary Mikhail Gorbachev and his wife, on her first visit to the United States. "Mrs. Gorbachev," the proto-col chief wrote later, "was certainly not an easy person. Although very bright, she had a flinty testiness and appeared overbearing." Raisa, with her university degree in philosophy, obviously did not wish to be equated with a former movie actress. Every detail of her schedule had to be negotiated, then renegotiated with Nancy. When they met again during the Reagans' visit to Moscow, the cold war had thawed measurably between the President and the General Sec-retary. But not between two very strong First Ladies.

IT IS A given that a President's wife, and even his extended family, must be wary of saying much of anything. More than any other First Lady, Betty Ford took exception to the gag order. When Betty talked, everybody listened—because you never knew what this can-did woman might say. A national outcry erupted in 1975 when she acknowledged on *60 Minutes* that her four teenage children had probably tried marijuana and that she would be unperturbed if her daughter told her she was having an affair, adding, "I would certainly counsel and advise her on the subject."

Conservatives and clergy thundered disapproval; liberated women cheered her. President Ford, loyal husband and loyal Repub-lican, could only squirm as his wife stepped boldly into a political mine field. "You just lost me twenty million votes," he groaned. As Betty laughingly told it years later, "He didn't kick me out of the house, but at one point he did throw a pillow at me." An unrepentant Betty elaborated, "My advice to anyone as First Lady would be to be herself. No one should have to live up to a standard." Thirty years later she sticks to her guns and continues to speak her mind. "I never thought being First Lady should prevent me from expressing my own opinions," she says. Bold words, but candor in the White House is not treated kindly by the public. Theirs was a short White House tenure, the remainder of Richard Nixon's second term, yet it was long enough for Betty Ford to become both a target and a force.

JACQUELINE KENNEDY, IN less than three years in the White House, created a mystique that has escalated into legend. Most people may not remember, or never knew, that she had been a lightning rod for criticism, a good deal of it from the West Wing. Too highfalutin, too self-centered, said her detractors—and maybe too young and beautiful. Hers was a remarkable metamorphosis from loose cannon in her husband's campaign to national icon following his assassination. More than ten years after her death and forty-odd years after she brightened the West's sky like planet Venus on a winter night, she still intrigues a generation not born when she stepped into the White House. To preservationists and history lovers, she was the magnet whose smile drew historic elegance into the White House and whose personal style stamped a generation.

She was also a headstrong First Lady who balked at White House restraints. She shirked official appointments and accepted costly gifts from foreign leaders. She cruised the Mediterranean on the spectacular yacht of Aristotle Onassis, her unseemly host who was—then—her sister's attentive friend and, more damaging politically, a shipping tycoon in litigation with the U.S. government. On a solo official visit to Greece in 1961, Jackie, in a mood described as "mean-spirited" by her staff director, Letitia Baldrige, was rudely dismissive of the official program tailored to her interests, heavy on art and history; Jackie preferred the yachts and parties of the jet-set. Watching the goodwill trip turn sour, Baldrige called the Oval Office to importune the President to save the mission. What JFK said is not known, but a chastised Jackie, ever the chameleon, sprinkled her fairy dust and dazzled the Greek prime minister—along with the European press—to make the visit a rousing success. She repeated that success in her colorful trip to India and Pakistan, with the leaders of the two hostile nations competing for her attentions.

With her arresting beauty and intellect, Jackie beguiled every head of state who crossed her path, from France's imperious de Gaulle to the Soviet Union's mercurial, rough-edged Nikita Khrushchev. Her grace and dignity in tragedy, her sense of history befitting the White House, put her beyond censure. Though she was often at odds with arbiters and advisers, she was at one with the President's House. And no one has done more to enhance it.

THE DAY THE little-known Illinois lawyer Abraham Lincoln won the Republican nomination in 1860, Mary Todd Lincoln, in a first for a candidate's wife, was interviewed by a journalist from the *Chicago Journal* and another from the *New York Evening Post*—and charmed both men. They pulled out all the adjectives. The *Journal* caller her "amiable and accomplished, a very handsome woman with a vivacious and graceful manner . . . often sparkling talker . . . who will do the honors at the White House with appropriate grace." Praising her "pattern of lady-like courtesy and polish," the *Evening Post* declared that "she converses with freedom and grace."

That would be the last unmixed praise Mary would ever hear; from that time she was relentlessly mauled by critics. Arriving in the White House, she was wrong-footed from the beginning, aspiring too obviously to be the queen of Washington society, insensitive to the deepening crisis of war. She made enemies more easily than friends, detesting some cabinet members, officious toward the staff who tried to protect her, and fiercely jealous of any woman who as much as talked with her husband—on one occasion making a humiliating public display. The President's aides called her "the hellcat." In even the most sympathetic view, Mary Todd Lincoln was a difficult woman.

Her Kentucky family, like the nation itself, was split between North and South; her brother, three half-brothers, and three brothers-in-law served in the Confederate army; four perished. Hated as a southerner, she was rumored to be a Confederate spy, an accusation that festered to the point that Lincoln felt obliged to take action. In a formal appearance before a congressional investigating committee he stated, "I, Abraham Lincoln, president of the United States, appear of my own volition before this committee of the Senate to say I, of my own knowledge, know that it is untrue that any of my family hold treasonable relations with the enemy."

There was no need to mention his wife's name; everybody knew. Mary declined all comment, insisting, "My character is wholly domestic and the public have nothing to do with it." She (and other presidential kinfolk) failed to understand that the public has everything to do with it.

Cracks were already beginning to show in Mary's armor. She plunged $29,000 into debt (a huge amount at that time), overspending wildly on elaborate gowns, buying a hundred pairs of gloves in three months. "I must dress in costly materials," she protested in her own defense. "The people scrutinize every article that I wear." She exceeded the $20,000 congressional budget for refurbishing the White House by $6,700. Even her forbearing husband was angry, chastising her for spending so extravagantly on "flub-dubs for this damned old house, when the soldiers cannot have blankets."

As the war widened and worsened in 1862, Mary gave one of her most lavish receptions, bringing in New York caterers and ordering elaborate decorations—the pièce de résistance was a huge model, concocted of sugar, of the warship *Union* bearing the Stars and Stripes. Abolitionists throughout the country assailed her for such insensitivity when Union soldiers were suffering in "cheerless bivouacs or comfortless hospitals," and a Philadelphia poet published a screed in verse entitled "The Queen Must Dance." That was but the beginning for the tragic mistress of the White House.

Not only her spendthrift style but her erratic behavior set people talking. Most shocking was the scene when she arrived late to join the President in his carriage to review the troops—and found that the commanding general's wife had taken her place. Furious, Mary exploded in an outburst that humiliated her husband and herself and set all of Washington gossiping. The White House staff removed the most vicious of the letters berating her.

Criticism and scorn followed her after her husband's death and her own mental disarray. Whatever she touched went wrong. Trying to raise money, she came up with a scheme to sell her White House gowns, incognito, in New York; of course, the dealer leaked the grieving First Lady's name. The usual brouhaha ensued, and poor Mary had to pay eight hundred dollars to recover her own belongings. She fled to Europe with her young son, Tad, moving from one second-rate hotel to the next, until the boy, then eighteen, developed tuberculosis. The pathetic pair returned to America, where he died, the fourth overwhelming grief for Mary to bear.

Her terrible tale did not end there. At crossed swords with her remaining son, Robert, who had never been close to his parents, she lost all reason and the son had his mother committed to an institution for the insane. At length she was released, made peace of a sort with Robert, moved in with her sister, and blessedly slipped away quietly. Mary Todd Lincoln was in reality a figure worthy of a Shakespearean tragedy.

When it comes to vituperation, Florence Harding could claim to have suffered the worst. Seven years after President Harding's death in California, with "the Duchess" at his side, Gaston B. Means, one of Harding's unsavory cronies and among several sent to prison, produced a book, *The Strange Death of President Harding*. It was a generally discredited tabloid-level effort, but the whole country was gossiping about his virtual accusation that the First Lady had poisoned her philandering husband. Florence, for reasons of her own, had retrieved much of her husband's correspondence and burned it all, which only served to feed the rumors.

After forty-six years in the White House, William Crook, first as Lincoln's bodyguard and then chief usher, reflected on the anguish the Executive Mansion can inflict on a President's family. He remembered Caroline Harrison reading the morning papers, weeping at the vicious ridicule against the President and the fact that "even his little grandchildren are made fun of." Crook's long experience led him to muse, "I do not wonder that many an able, brilliant man refuses to enter public life simply because he will not subject himself and his family to such misery." Since 1865, when Colonel Crook's White House career began, White House families would say not much has changed.

IT'S NOT JUST First Ladies who run into trouble in the White House. Presidents accustomed to political attacks in the Oval Office can find themselves caught in unexpected fire in what is supposed to be the tranquil side of the White House. The complaints have usually been picky, like Jefferson's decision to abolish the weekly White House receptions. John Tyler was chastised for serving poor meals and presiding over what the press dubbed "the Public Shabby House." John Quincy Adams, the embodiment of rectitude, got into trouble for such a trivial offense as charging a billiards table to the White House, an issue settled only when his son paid the $84.50 out of his own pocket. (Consider that the White House now includes a workout room among its many amenities—and that Laura Bush raised the bar for First Ladies by being the first to lift weights.)

The most egregious case of a President plunging into hot water set the whole country talking. One morning in 1950 *Washington Post* music critic Paul Hume was handed a letter from the White House. He must have trembled with excitement; if not, he should have, because President Truman was on a rampage. His daughter, Margaret, an ambitious young soprano, had performed the previous evening before an audience of thirty-five hundred in Washington's foremost concert hall. Hume, after describing her as "extremely attractive on stage," turned to her singing: "Miss Truman cannot sing very well. She is flat a good deal of the time. . . . There are few moments when one can relax and feel confident that she will make her goal, which is the end of the song. . . . She still cannot sing with anything approaching professional finish. . . . And still the public goes and pays the same price it would for the world's finest singers."

Reading the review, the President exploded, firing off a handwritten note on a White House memo pad: "Mr. Hume: I've just read your lousy review of Margaret's concert. I've come to the conclusion that you are an 'eight-ulcer man on four-ulcer pay.' It seems to me that you are a frustrated old man [Hume was thirty-five] who wishes he could have been successful." His critique of Hume's critique: "poppy-cock." "Some day I hope to meet you," Truman growled. "When that happens you'll need a new nose, a lot of beef-steak for black eyes, and perhaps a supporter below!"

In defending his daughter, the devoted father failed to consider the consequences. Margaret and Bess were mortified. Truman clung to his right as a father—who just happened to live in the White House—to defend her. The reality is that the man in the White House is President at all times, and this President's diatribe, which made headlines around the world, was out of line. Truman was confident that the public, as parents, would back him, but his instincts were dead wrong—White House mail ran two to one against him, outraged by his vulgar language and loss of self-control. The presidency transcends a father's personal emotion.

A dispassionate look at the dustup suggests that the Trumans—parents and daughter—brought this embarrassment on themselves. Margaret, then twenty-six, had forgotten her own self-warning about the danger of being loved for your White House connection rather than yourself. Her New York voice teachers had told her she was not yet concert-ready, and world-famous opera star Helen Traubel, a friend, had said the same to both Margaret and the President, with the warning that she would be seen as riding on his position. Only Margaret's teacher back in Independence assured her that she was ready for the concert stage—and that was the opinion Margaret chose to follow.

It was easy—no surprise—for the President's daughter to start at the top, on the nationally broadcast *Sunday Evening Hour*. Her appearance drew its largest-ever number of listeners; at her concert in the Hollywood Bowl, she received seven curtain calls; in Pittsburgh, nine curtain calls and three encores. She was praised everywhere for her poise and charm—less so for her talent.

Years later Margaret Truman wrote in her memoirs, "Perhaps sheer naivete saw me through. I was probably the first unevaluated singer to make a debut with a major symphony to an audience of twenty million, a gamble which would have turned a more sophisticated singer gray." (But she still seemed to miss the importance of her White House connection.) Margaret, always attractive and likeable, became a television personality and a writer of proven talent.

Harry Truman was not the only protective President-father. For Washington reporters the Theodore Roosevelt years were a field day; the extrovert President was both accessible and agreeable, and his mischievous children were always good copy. TR was a master of "spin," and on more than one occasion he wrote out in his own hand exactly what he wanted to be sent over the press wires that day.

It was a cozy (and professionally deplorable) relationship, until a newspaperman, obviously hard up for a story, reported that the Roosevelt children had chased and frightened—perhaps tormented—several turkeys that had the run of the White House grounds before Thanksgiving. The President, insisting that the story was inaccurate, issued a fatwa against the hapless reporter, denying him not only White House access but all government press releases. The ban was, of course, impossible—not to mention ridiculous and probably illegal—but a chief executive who coolly handles attacks on his policies becomes a hotly defensive father when his children are targeted.

WHEN SMEARS, EXAGGERATED or totally false and always politically inspired, are circulated about a President's personal life, he faces a quandary: deny and give the fabrications wider circulation, or stay silent and risk seeming to accept guilt? Theodore Roosevelt was the one President who roared back, suing the newspaper that intimated he was a drunkard. He explained his predicament in a public statement: "Any man familiar with public life realizes the foul gossip that ripples beneath the surface . . . especially about every President. It is only occasionally printed in reputable papers . . . but it is hinted in private. And if it is left unrefuted, after a man's death it lasts as a stain which is then too late to remove. From Lincoln to Garfield to Cleveland and McKinley, this gossip has circulated and still circulates."

The Cleveland gossip might have referred to the illegitimate child Cleveland never denied fathering in his extended bachelorhood. Or it might have referred to the rumors that circulated during his bid for reelection in 1888. The twenty-second President, proud of his young and beautiful wife and reveling in belated domesticity, found himself the ultimate victim of "when did you stop beating your wife?" It was whispered that the President, in a drunken rage, had assaulted his popular young Frances and locked her out of the White House.

Frances Cleveland did what no other First Lady had ever dared to do: she fought back. In a statement she released to the press, the First Lady declared, "I could wish the women of our country no greater blessing than that their homes and lives may be as happy, and their husbands as kind, attentive, considerate and affectionate as mine." Long years of what by all accounts was a loving marriage free of any further rumors—in their idyllic post–White House life with their five young children, they were the most prominent family in Princeton, New Jersey—suggest that the rumor had been campaign politics at its most execrable.

Franklin Roosevelt, the perpetual target of isolationists and big business, ran into an ambush that remains unique. In the summer of 1944, returning from a meeting with his Pacific commanders in Pearl Harbor, FDR faced a Republican tsunami. His foes spread the story that his beloved Scottish terrier, Fala, had somehow been left on an Aleutian island and that FDR, as commander in chief, ordered a destroyer—at great expense—to fetch him. The facts of the tale are murky at best, but the President chose to make the most of it: "I don't resent attacks, and my family doesn't resent attacks," he declared in mock indignation, "but Fala *does* resent them. Fala is Scotch [and] his Scotch soul was furious. He has not been the same dog since." His audiences loved it—in those days the country was made up of more dog lovers than Republicans.

Politicians have to be careful how they deal with dogs, who can always upstage them. Twenty years after the Fala expedition, Lyndon Johnson unleashed an attack when he picked up Little Beagle Johnson by his floppy ears—in full view of press corps and cameras. Across the country the dog constituency barked its outrage. Despite LBJ's insistence that it didn't hurt, Little Beagle Johnson's yelps suggested that he had not been polled.

ONE SURE WAY to cause an uproar is to try to fancy up the White House; Americans have an inborn resistance to too many frills, as Martin Van Buren learned when he sought reelection in the midst of a deep depression in 1840. As Vice President, Van Buren had ridden into the White House on Andrew Jackson's long coattails, but his own sumptuous White House lifestyle was seized by his Whig opponents as the issue separating the effete East from the virile West. This led to a campaign speech unequaled in both length and adjectives: for three straight days Pennsylvania Congressman Charles Ogle excoriated Van Buren on the floor of the House of Representatives.

Ogle was a veritable thesaurus of extravagance. Journalist Perley Poore reported the marathon polemics with amusement: "He dwelt on the gorgeous splendor of the damask window curtains, the dazzling magnificence of the large mirrors and chandeliers . . . the satin-covered chairs . . . the imperial carpets and, above all, the service of silver, including a set of what he called gold spoons, although they were silver-gilt."

Contrasting Van Buren, a wealthy Hudson Valley New Yorker, and his effete luxuries with the Whig candidate, Ogle limned General William Henry Harrison as a rugged man of the West, a general with a taste for hard cider, a coonskin cap, and a log cabin decorated with only "a string of speckled birds' eggs . . . and a fringed window curtain of white cotton cloth." "I put it to the free citizens of this country," the congressman thundered. "Will they longer feel inclined to support their chief servant in a palace as splendid as that of the Caesars and as richly adorned as the proudest Asiatic mansion?" Never mind that Harrison grew up at Berkeley, still standing as one of the finest of Virginia's James River plantations. It was visited on occasion by George Washington, a friend of Harrison's father's, who was a member of the Continental Congress, a signer of the Declaration of Independence, and a three-time governor of Virginia. Even in 1840 image was everything.

Winding up his three-day oration, Ogle pulled out the most scandalous charge of all: that Van Buren, this man seeking another term, spends his time lolling about in a tepid bath, spraying his whiskers with *French Triple Distille Savon Davelline Mons Sens.* Tepid bath! French whatever! Running without Old Hickory's coattails, Van Buren was defeated in a rout.

"HOW YOU GONNA keep 'em down on the farm after they've seen Paree?" went a favorite song from the World War I era, and the words are still relevant in the White House. Presidents and First

Ladies travel in the personal perfection of *Air Force One*, a virtual flying hotel; they are fêted at the grandest of palaces and are shown the most awesome treasures. (To be sure that the visiting American leader is not exposed to scenes of poverty and deprivation, some leaders—egregiously, Imelda and Ferdinand Marcos during their reign in the Philippines—have actually erected false fronts along the route, a literal Potemkin's village.

A foreign visit sounds glamorous, and often it is—the Reagans spending a weekend with the Queen at Windsor Castle, where the two heads of state gracefully ride her finest Thoroughbreds; the LBJs greeted with ropes of flowers and beautiful native women in sarongs chanting polynesian songs on the Gauguinesque island of Pago Pago; Jackie Kennedy riding in a howdah atop an elephant at a rajah's palace in India. But there is a downside, too: jet lag and strained small talk through an interpreter's eccentric English; motorcades greeted by protest signs (paint was hurled at LBJ in Australia); food sliding off the silver chopsticks in South Korea's equivalent of the White House; too many speeches and too little sleep.

State dinners can be memorable. Hillary Clinton tells of Boris Yeltsin's fête in an ornate dining hall in the ancient Kremlin. "I have a very special treat for you tonight!" her host confided. "What is it?" "You must wait until it comes." Then he relented: "Moose lips." Hillary described her reaction: "Sure enough, floating in the murky broth was my own set of moose lips. The gelatinous shapes looked like rubber bands that had lost their stretch, and I pushed them around the bowl until the waiter took them away. I tasted a lot of unusual food for my country, but I drew the line at moose lips."

A president often looks around at the splendid cities he visits and thinks, Why don't we dress up our capital a bit? Sometimes that's a good thing, as when President Kennedy and Jackie, returning from Paris and Vienna, took a hard look at tatty Pennsylvania Avenue, the broad street linking the Capitol to the White House, and decided something had to be done about it. With Kennedy's nudging, the street was transformed over time to fulfill Pierre L'Enfant's vision when he laid out the new city in March 1791—a handsome avenue lined with institutions that would be "attractive to the learned and afford diversion to the idle." And every four years worthy of an inaugural parade.

Whenever foreign ideas are borrowed for the American capital, there is the risk that something may be lost in translation. President Nixon, on his first visit to Europe as President, was impressed by the splendid uniforms worn by the various palace guards and decided to smarten up the admittedly undistinguished White House counterparts. New uniforms were designed; routine police blue gave way to white tunics bedecked with gold braid; standard police caps were replaced by high-peaked headgear. The guards wore the new outfit with the pained look of dogs dressed up in kids' clothes. The media had a field day, and metaphors ran wild— the new uniform was inspired by a Ruritanian customs official; it was borrowed from a Viennese operetta company. It would have caused Thomas Jefferson, the apostle of republican simplicity, to have one of his migraines. The White House triumphed on grounds of dignity; the President had to concede. Somewhere—so it was said later—a high school band paraded proudly in the guards' scarcely used White House uniforms.

A gloss of added elegance can be a good thing, like Jackie Kennedy's goal to bring historic authenticity to the White House, or not so good, as with Caroline Harrison's overblown plan for rebuilding it as a veritable palace or Nellie Taft's royalist "improvements." Wielding her position like a newly crowned queen, Nellie summarily replaced loyal, long-serving ushers in their traditional frock coats with six black footmen in blue livery. Quite pleased with their appearance, she declared, "They lend a certain air of formal dignity to the entrance, which, in my opinion, it has always lacked."

The public and the White House reacted with ridicule. Alice Roosevelt Longworth commented dismissively, "Mrs. Taft had her own ideas of how the White House should be run. They were rather grander than ours." Nellie attributed the derision to "adherents, sincere and otherwise, of our too widely vaunted 'democratic simplicity.'" Four years later, following the defeat of President Taft—and Nellie—the ushers were back in their traditional black frock coats; the White House returned again to Thomas Jefferson's "democratic simplicity."

Again and again the White House has resisted attempts to grandify its classic restraint. From the earliest days it shunned the fancy uniforms cherished by noble or ostentatiously rich households. Over the years there have been some close calls, but at each moment of peril its residents with the cultivated taste of Edith Roosevelt and Jackie Kennedy or the historic passion of Harry Truman—and concerned citizens on the outside—have rallied to stave off ill-conceived or irreparable "improvements."

The solution to the necessity of barricading the White House against attack was handsomely revised in 2004, making it not grander or more forbidding, but more visitor friendly. Brutal cement barriers and daunting checkpoints were replaced in 2004 by huge planters filled with seasonal flowers, serving the same purpose as the forbidding barriers. Pennsylvania Avenue in front of the White House, which retains its high fence of iron palings, remains permanently closed to vehicles, and other security measures are in place, but now pedestrians can stroll the new brick plaza or rest on its benches (another attractive bit of subtle security), and in tree-shaded Lafayette Park office workers can bring their lunch or take the sun. Towering American elms—John Adams's favorite tree—line the two blocks incorporating Blair House and the wedding cake structure built in the Grant era to house the State, War, and Navy Departments, which has accommodated overflow from the White House offices since World War II. The new layout, enthusiastically dedicated by Laura Bush, provides security with charm. It has returned a pleasant civility to all who want to enjoy the environs of their White House. Once again the President's House has accommodated changing times. Abigail Adams wouldn't recognize the "great castle" today, but she was right: it has indeed been "capable of improvement" and "built for the ages to come."

twelve

Dark Days

————◦————

FROM ITS BEGINNING the White House has been at the center of the nation's great moments, events that range from triumphant to tragic—and, for Presidents and their families, times that have been exhilarating or humiliating. FDR did not live to see America dancing in the streets at the final victory of World War II; Lyndon Johnson, lost in the political jungles of Vietnam, lived with the cacophony of hatred outside his windows. Edith Galt Wilson went from sparkling First Lady to duplicitous wife who misled the Congress and the nation in her determination to keep her incapacitated husband in the White House.

Three times the unthinkable specter of impeachment faced a President. A one-vote margin in the Senate saved Andrew Johnson from being thrown out of office; Bill Clinton withstood a similar challenge. For their families, however, there was a vast difference—Johnson was the political target of powerful members of the Lincoln cabinet who resented and despised him; Clinton's sordid toying with an aggrandizing young intern led to public humiliation for his wife and daughter. Richard Nixon, facing impeachment on charges arising from the Watergate affair, was advised that he would be found guilty by the Senate; he resigned rather than become the only American President to be removed from office.

After leaving office, all three led productive lives. Thousands gathered on the White House lawn to cheer Andrew Johnson's acquittal, and White House social life flourished. The throng of well-wishers at his farewell reception was the largest the old mansion had ever witnessed. After several years Nixon was informally restored to the role he most enjoyed, elder statesman in foreign policy. Bill Clinton left office with high ratings in the polls and remains one of the country's most popular political figures.

MOST WRENCHING OF all the White House tragedies is the loss of a child, experienced first when the Lincolns' eleven-year-old Willie, their pride and joy, was lost to typhoid fever. Mary's collapse in grief left Lincoln to bear his anguish alone, adding to the heavy burden of the war. In times of great personal distress—and there have been many—the White House cossets its stricken families, but when they most wish to grieve in solitude, it cannot shut out the world.

Twenty years earlier, at the death of President Tyler's wife—the first White House death—his daughter-in-law, Priscilla, wrote to her sisters: "Nothing can exceed the loneliness of this large and gloomy mansion—hung with black—its walls echoing our sighs." That loneliness is magnified when it marks the death of a child.

It would be another sixty-two years before the unbearable loss of a President's child struck again. The Coolidge White House instantly became a livelier place when their two teenage sons were home for their summer vacation from Mercersburg Academy in Pennsylvania. June 1924 was special: John had graduated and was soon to enter Amherst College, which meant the two brothers, born only eighteen months apart, would be going their separate ways for the first time. Sixteen-year-old Calvin, wearing no socks in a morning of tennis on the White House courts, rubbed a blister on his foot. It was nothing really, just a blister—until it turned lethal and a staphylococcus infection (then known as blood poisoning) set in.

His distraught father, going through the motions of living,

caught one of the many rabbits that roamed the White House lawns and, cradling it in his arms, took it to Calvin's room in the faint hope of amusing him. It was a touching effort by the President who had always seemed bloodless, but there was no hope; lifesaving antibiotics were more than a decade away. The son who was outgoing and full of fun, so unlike his father, lost the battle. As his son's casket was borne from the White House after the funeral, the dour President wept. In his memoirs Coolidge put words to his grief: "When he died, the honor and glory of the Presidency went with him." Grace Coolidge later expressed her thoughts in a poem, published in *Good Housekeeping*, that began:

> *You, my son*
> *Have shown me God*
> *Your kiss upon my cheek*
> *Has made me feel the gentle touch*
> *Of Him who leads us on.*
> *The memory of your smile, when young,*
> *Reveals His face*
> *As mellowing years come on apace . . .*

It was August 1963, and the White House was in a flurry of anticipation, readying the nursery for the newest Kennedy. Jackie, who had a history of unsuccessful pregnancies, was taking it easy on Cape Cod. Suddenly everything went wrong. The President rushed from the White House to Otis Air Force Base hospital, where his baby son—born premature, weighing less than five pounds, and diagnosed with hyaline membrane disease—was clinging tenuously to life. The hospital chaplain hastily christened him Patrick Bouvier Kennedy, the name his parents had chosen, and he was rushed to Children's Hospital in Boston.

All the while, the President was keeping the grave concern from Jackie. As the infant's chances faded, he sat all night beside the hospital bassinet, holding his son's tiny fingers in his own—until Patrick gave up his fragile grasp of a life that had flickered for a single day. The President of the United States was helpless. Returning to his wife's bedside, he sank to his knees and sobbed. Boston's Cardinal Cushing, who officiated at the Mass of the Angels in his private chapel, later recounted that after the others had left the chapel, Kennedy "put his arms around the casket of little Patrick, and I can tell you in all honesty he tried to take it with him. And he wept." Back at the White House, the staff, in tears, dismantled the nursery that had been waiting for baby Patrick.

The deaths of three children in the White House over more than two hundred years give a false impression of the scope of childhood mortality. The dark demographics of the nineteenth century show how painfully normal it was for a family to lose a child to diseases that brought dread and death. (In 1800 American life expectancy was 35.2 years, but if the most vulnerable could make it past their teen years, they had a good chance of reaching the average life span of 55.3 years.) The diphtheria, scarlet fever, whooping cough, and septicemia that snuffed out the lives of the very young have been virtually stamped out by antibiotics and other modern medicines; malaria and typhoid and typhus fevers have been largely brought under control by both medicine and public health measures.

But thirty-two presidential children have died before they were out of their teens, twenty-seven of them in the nineteenth century. With the exception of the Lincoln's Willie, all died before their father was President. The Jeffersons lost four of their six before the age of two and a half; the Pierces grieved for their three sons, one killed in an accident; the McKinleys buried both their little daughters. In the twentieth century the death of the infant Franklin Roosevelt, Jr., led the parents to bestow the name again on the next boy; the Eisenhowers lost their first son to scarlet fever when he was only three. The most contemporary death was that of George and Barbara Bush's little Robin, who died of leukemia in 1953, before she was four.

In the nineteenth century, when the life span of women was shortened by multiple childbirths and illnesses, four Presidents were widowers. Two First Ladies, Letitia Tyler and Caroline Harrison, died in the White House. Ellen Wilson was the only loss in the twentieth century.

EIGHT TIMES THE White House has been shrouded in the black crape of mourning for a President: four have died from natural causes, four from an assassin's bullet. The first loss began on a raw March Inauguration Day with a bitter wind slashing across the Capitol Plaza. General William Henry Harrison, hero of the Indian wars and, at sixty-eight, then the oldest man ever elected President (only Reagan, at seventy-three, has overtaken him), was showing the stuff he was made of. He declined the fine carriage presented by the Whigs of Baltimore—ever the general, he insisted on riding his white charger to the Capitol, lifting his hat to acknowledge the cheering throngs along the way.

"He stood bareheaded, without overcoat or gloves, facing the cold northeast wind," wrote journalist Perley Poore, who covered the event. Dignitaries on the platform, warmly wrapped, "suffered from the piercing blasts." Harrison had dismissed his advisers' tactful suggestion that he shorten his speech, which stretched for 8,578 words, including forty-five uses of the vertical pronoun. He spoke for one hour and forty-five minutes, still the record for an inaugural address—one that we hope will not be challenged. His wife, Anna Symmes Harrison, had remained at their home in Ohio, recuperating from an illness and preparing for the new life she dreaded, the public life of a First Lady.

The vigorous old Tippecanoe reviewed a long and tumultuous parade, held a large reception during the afternoon, and attended three inaugural balls. His two sons and other relatives were staying in the White House, and even the state banquet table could not accommodate the forty friends who came for dinner one evening. Feeling quite at home in the Executive Mansion, though surely exhausted from such activity, Harrison more than once went shopping at Center Market. (Imagine a President today shopping for his own table, unaccompanied.)

One morning, going from stall to stall with a basket over his arm and as usual wearing no overcoat, "he was overtaken by a slight shower and got wet," Poore reported, "but he refused to change his clothes." Within hours he developed a heavy cold and fever, which worsened into pneumonia before doctors were summoned. "His physical powers, enfeebled by age, had been overtaxed," Poore suggested. For a week four doctors employed every known remedy—bleeding, dosing, cupping, blistering. (Today's specialists conjecture that his death may have been caused by the extreme treatments rather than pneumonia, an irony, of course, and even more so as Harrison was the only American President to have studied medicine.) Finally the old general, the new President, could fight no more. On April 4, 1841, exactly one month after he was sworn in, the rugged old soldier the Indians could not defeat was felled by Washington's weather.

For the first of what would be too many times, the White House had to deal with the death of a President, setting an elaborate pattern that has been followed, with minor changes, ever since. The East Room was put to a sorrowful new use. The coffin was placed on a catafalque draped in black velvet trimmed with gold lace. Over that was a velvet pall with deep golden fringe, bearing the Sword of Justice and the Sword of State, and atop it all was the scroll of the Constitution. The funeral car was pulled by six white horses, each with a black groom dressed in white, with white turban and sash. The procession stretched for two miles.

Far away in Ohio, the wife who had bitterly opposed his running for President, the First Lady who would never so much as glimpse the inside of the White House, could only grieve, adding this new blow to the untimely deaths of eight of their ten children. Only two Harrison offspring lived beyond a thirty-fifth birthday, the eldest daughter and a middle son, John Scott Harrison, a two-term congressman who became better known as the father of Benjamin Harrison, the twenty-third President.

Despite his place in the Harrison dynasty, Benjamin, the only President to follow in the footsteps of a President-grandfather, hated life in the White House. He complained that having the President's office in the heart of the family quarters was "an evil combination" and shared President Fillmore's view that the mansion was "a temple of inconvenience."

His wife, the musical and artistic Caroline, was an active First Lady after shelving her plans for rebuilding the White House, but her severe husband, dubbed "the Ice Box," was simply not suited to the job. When a friend said to him in exasperation, "For God's sake, be human!" Harrison replied, "I tried it and I failed. I never tried to be human again." If that was a wry joke, it contained the essence of truth. "My life," he once remarked, "is about as barren of anything fun as a desert is of a blade of grass." The late historian Richard Neustadt rated him "the coldest fish we ever had as President." Passive and negative are not the description of a successful President.

In the last months of the 1888 campaign, Benjamin's beloved Caroline fell ill of a fever thought to be typhoid; the President was at her bedside throughout the nights, his eyes red with weeping. Two weeks before the election, she died. His defeat paled beside this tragedy. "After the heavy blow of my wife's death," he said, "I do not believe I could have been President."

Nine years later the White House was once again draped in black for Zachary Taylor, another old soldier, the general credited with defeating Santa Anna in the War with Mexico. It was a broiling Fourth of July at the Washington Monument, where a large crowd gathered for an interminable amount of speech making, winding up with the popular President. Back at the White House, the chief executive ate his way through the better part of a basket of cherries, washed down with several tall glasses of iced milk, and after dinner he had another round of cherries and iced milk reportedly over the protests of his physician, who was his guest.

Within hours the president fell ill with a little stomach problem. For five days his condition steadily worsened; on the third day he told his doctor, "In two days I shall be a dead man." His self-diagnosis proved all too accurate: on the fifth day, surrounded by his family and top officials, he murmured, "I have endeavored to do my duty," and peacefully breathed his last. The bells tolled at midnight. Old Rough and Ready was gone; Millard Fillmore became America's thirteenth President. Taylor's death was attributed to various causes—a stomach infection, cholera morbus, gastroenteritis. And there was talk of poison, enough that in 1991, almost 150 years after his death, his body was exhumed to settle the lingering question. No trace of arsenic was found, leaving cholera the likeliest culprit for having broken the ordered sequence of presidential succession.

ON APRIL 15, 1945, the *Ferdinand Magellan*, the specially fitted, armor-plated presidential train, was making one last journey to Hyde Park, taking the body of Franklin Delano Roosevelt back to his beloved home on the forested slopes of the Hudson. Its seventeen cars were filled with men and women who were his friends and had worked with him for years, usually a talkative crowd but subdued that day, still unable to comprehend the sudden loss of their old friend and world leader. In the brief East Room service, Episcopal Bishop Angus Dun had sought to rally spirits by quoting the valiant message the yet-untested FDR had unfurled like a banner at his first inauguration—"The only thing we have to fear is fear itself." The memory of his voice, his valor, brought tears to the eyes.

The train proceeded north at respectful funeral speed. Throughout the night, in its seventeen sleeping cars, the mourners were lost in their own thoughts. In the choicest compartment the new President,

Harry Truman, knew that many on the train viewed him as an unin-
tended consequence. Sleepless in another car, Henry Wallace, FDR's
third-term Vice President who was dropped in favor of Truman,
had time to think—who knows what? But for one decision—Roosevelt
needed support in the Congress, and the leaders on the Hill found
Wallace's politics too far left—he could be on this train as President.
He looked out into the night that was taking him down a different road
and later wrote: "Although there weren't many stars out and no visible
moon, I could see the silent, bowed crowds that lined the sorrowful
tracks to pay tribute to their fallen leader. Everything seemed dirgelike.
Even the wheels sounded like so many muffled drums."

Bess Truman, also lying awake, was overcome by the President's
death—she still thought of Roosevelt as the President—and was
panicked by the upheaval in her own life. "I was very apprehensive,"
she later recalled. "The country was used to Eleanor Roosevelt. I
couldn't possibly be anything like her—I wasn't going down in any
coal mines."

Back in the Executive Mansion, a somber crew worked night and
day for a week, packing twelve years of the Roosevelts' accumulated
possessions into twenty large army trucks. Men and women whose
lives had been entwined with the Roosevelts for so many years were
anxious, bereft, unmoored. The world had been shaken in the midst
of the greatest war, and now their personal world had been sundered. The day after Eleanor Roosevelt left, Chief Usher West remembered, "the White House was like a ghost house," shabby and
faded "like an abandoned hotel."

ON APRIL II, 1865, two days after General Lee surrendered to General Grant, an exuberant crowd converged on the front lawn of the
White House. "The public heart began, almost too suddenly for re-
alization, to breathe freer and happier in the anticipation of speedy
peace," wrote a Lincoln contemporary of the Civil War's end. "It was

but natural that the people should instinctively desire to congratulate their great leader . . . now relaxed and beaming with a grateful sense of relief from anxiety."

At the White House, its lights ablaze, they called for the President. He stepped to a window overlooking the North Portico and began to speak: "We meet this evening, not in sorrow, but in gladness of heart." He finished to great cheers and patriotic tunes by the Marine Band. Never had the White House been the scene of such exhilaration, such triumph. Among the crowd was a young actor named John Wilkes Booth.

On April 14, as part of his busy day, Lincoln held a cabinet meeting, one of the most important of his presidency, which produced agreement on his postwar policy for the South, restoring the southern states "to their ancient place beside their sister states." After a pleasant afternoon carriage ride, First Lady Mary, released at last from the crushing anxieties of the war, was free to laugh at *Our American Cousin* at Ford's Theater. Knowing that all eyes would be on the President's box at their first social appearance since the surrender, she selected an elegant dress from among her many. Washington was rejoicing, ready to live a full life again, and Mary, who loved her role as First Lady, was ready to lead the way in the White House.

Leaving his office, the President told his personal aide, William Crook, about the disturbing dream that had come to him on the past three nights; in the dream he was assassinated and he saw himself lying in a coffin. Crook later said he pleaded with the President not to go out, but Lincoln replied in a husbandly way that he didn't want to disappoint his wife. Then, the aide recalled, instead of his customary "Good evening, Crook," he said, for the first time ever, "Good-bye, Crook."

It is gripping to read in an 1865 volume by L. P. Brockett, a Lincoln friend, the depositions of those who were with the President in Ford's Theater, their immediate statements, not affected by hindsight:

John Wilkes Booth enters the presidential box, fires his pistol point-blank at the President, slashes Major Henry Rathbone, who was attempting to seize him, with a large knife, and leaps onto the stage twelve feet below, shouting *"Sic semper tyrannis!"* Mary screams, then faints. "All was instantly confusion." Doctors who rushed to the box at first could not locate the wound; they then quickly moved the President, unconscious, to a private home across the street.

The following morning, Mary, sobbing uncontrollably, was prostrate on her dying husband's body. As she realized he was gone, she screamed and fainted—whereupon Secretary of War Edwin Stanton, her greatest enemy, loudly ordered, "Take that woman out of here and do not let her in again!" It is painful to read such cruel words aimed at a wife in incomprehensible agony. Her son, Captain Robert Lincoln, took her back to the White House—a bitter journey of a few blocks to an unendurable life ahead.

The nation—the world—was in shock. The White House had lost a President before, but the very word *assassination* was unthinkable. The war had at last ended, and now the leader, the Great Emancipator, the healer, was gone. Within minutes, Andrew Johnson, Lincoln's compromise choice for Vice President, a man from the Tennessee mountains who deplored secession as "hell-born and hell-bound," was President.

The murdered President lay in state in the East Room with soldiers standing guard at his catafalque (which, almost one hundred years later, would bear the coffin of President Kennedy). Chandeliers and windows were draped in black and the great mirrors were masked in white muslin. The distraught First Lady did not leave her room, nor did she accompany the body on the funeral train that bore him back to Illinois, a twelve-day journey of 1,700 miles, with stops for sorrowful ceremonies and an open coffin in eleven cities along the way. The remains of his beloved Willie joined his father's body on the long trip home. The cortege, wrote L. P. Brockett, who witnessed the events, "met with such a spontaneous and magnificent reception as had never, on this continent, been hitherto accorded to any man living or dead."

Today, in close contact with history, audiences at Ford's Theater can look up to the flag-draped presidential box and in a small museum on the lower floor see the overcoat Lincoln was wearing—its left sleeve ripped off by a macabre memorabilia hunter. Across Twelfth Street, the house the dying President was taken to still stands, as a museum now, with Lincoln's death room as it was that mournful night.

LINCOLN COULD BE called the final casualty of the Civil War. But how can the nation grasp the killing of popular, noncontroversial presidents by deranged nobodies like those who assassinated Garfield, McKinley, and Kennedy?

July 2, 1881, began as a fine day for James Garfield, a well-liked President just four months into his term. His cherished helpmeet, Lucretia, had at last recovered from malaria; his political conflicts had been put to rest; his family was looking forward to a summer holiday in the house he had rented on the New Jersey shore. As journalist Perley Poore put it, "Thus gladsome . . . strong, healthy and happy," the President was on his way to deliver the commencement address at his alma mater, Williams College, in Williamstown, Massachusetts.

As he entered the Baltimore and Potomac Railway station, two shots rang out. The President crumpled to the floor. One shot missed; the other pierced his spine. A deranged thirty-six-year-old transient, Charles Guiteau, who had foolishly fancied himself a U.S. consul abroad, had vindicated his rejection by shooting the President.

At the White House the doctors determined that the President's wounds were not fatal and began issuing daily optimistic reports. "I should think the people would be tired of having me dished up to them in this manner," he protested. Inventor Alexander Graham

Bell rushed his newest invention to the Executive Mansion, a metal detector that would pinpoint the bullet for removal, with explicit instructions to keep it away from all metal. But the doctors fussing over the President ignored the order and kept him on the bed's steel springs—thus neutralizing the device. Might it have worked?

To speed his recovery, Garfield was transported to Elberon, New Jersey, where rail tracks had been laid to the doorstep of his ocean-front vacation house. Over a period of eighty days his mind and body deteriorated together, his condition sinking from troubling to terminal. On September 19 the twentieth President slipped into history. Blood poisoning, not a bullet, caused his death. Lucretia, frail and distraught, wept, "Oh, why am I made to suffer this cruel wrong!" After less than twenty-nine weeks in the White House, her life's love and their dreams were gone.

A tattered copy of the *Hartford Evening Post* of September 20, 1881, conveys the national grief. Heavy black rules separate the columns, and the report is written in doleful cadences: "The long suspense is ended, and this good man who was our President is dead. . . . His death is an intimate personal grief to the fifty millions of his countrymen. . . . Hail and Farewell, oh Brother of our hearts!" The American President's death reverberated around the world. From London Queen Victoria, who still mourned her irreplaceable consort, Albert, sent condolences to Lucretia: "May God support and comfort you as He alone can."

A *Hartford Post* reporter rang the bell at the New York town house of Chester A. Arthur, the new President, and inquired if there would be a comment. The butler replied, "I dare not ask him. He is sitting alone in his room sobbing like a child, with his head on his desk, and his face buried in his hands. I dare not disturb him."

THE TRAGEDY OF John F. Kennedy still sears the memories of millions of Americans who were there, through television, in the reality of his murder, from the moment the shots found their target in his open limousine, through four days of mournful ceremony, to the notes of "Taps" floating like silver tears over the grave at Arlington Cemetery. With haunted eyes and stoic grace, the young widow kept her children at her side at St. Matthew's Cathedral. A silent crowd, tears streaming, witnessed the indelible moments of John junior's heartbreaking salute to his father, and Jackie's resolute walk to the White House, along with the assemblage of world leaders, behind the horse-drawn casket. Her sense of history in every detail dominated her personal despair and elevated her to a unique position among First Ladies. (None of the other three First Ladies whose husbands were assassinated attended their funerals.) Forty years later, in 2003, the outpouring of remembrance in the media and the tide of books marking the anniversary of John Kennedy's death was unprecedented.

One small, and surprising, White House event was little noted at the time. In the midst of her grief Jackie went ahead with the plans for little John's third birthday party on November 25. She insisted that her son, the bearer of the name, would leave the White House with a happy memory. Many years later, as she faced her own death, Jackie, in her orderly way, wrote final notes to her children. To John, who was so clearly marked for a shining future, she had a last motherly reminder: "You, especially, have a place in history . . ."

ON MARCH 30, 1981, Nancy Reagan emerged from George Washington University Hospital carefully composed, her calm demeanor the onetime actress's finest performance. Only hours before, her husband had been struck by another of those unbalanced ciphers who have sought their moment of fame by targeting American presidents. Reagan's jaunty comments as the surgeons prepared to operate to remove the would-be assassin's bullet had reassured the media and the nation; Nancy would not allow her face to reveal that, to the contrary, her husband's life hung in the balance. Long afterward, the former First Lady gave her real story of "the worst, the worst" day: "I almost lost him. People didn't know that—they only knew the things he said."

Reagan beat the odds and went about reshaping the political world for another seven years, bright years he shared with Nancy and almost no other. Ten years later, still jaunty in public, he told the nation that the dark curtain of Alzheimer's was drawing around him. Through the ever-darkening decade that took him further and further away from her, Nancy continued to do what she had always done—live her life for Ronnie, even when he no longer knew who the pretty lady was. And then, frail and spent, she attended him with grace at the state funeral that sealed his place in history.

In the grim pattern of White House statistics, death had stalked the presidents elected at twenty-year intervals beginning with William Henry Harrison in 1840, proceeding in almost ghoulish order to Lincoln in 1860, Garfield in 1880, McKinley in 1900, Harding in 1920, FDR in 1940, and JFK in 1960. Ronald Reagan, elected in 1980, had broken the jinx; he completed two full terms—but his doctors later revealed that the bullet was only an inch away from adding another president to that eerie list.

thirteen

Bittersweet Farewells

———◆———

IT WAS A Sunday, the day before the inauguration of 1969. Lady Bird Johnson was making her "final assault" on packing, facing the "shambles" of her beautiful room. "I remember standing by the window, taking a last look from that spot at the greatest view in Washington, out over the South Lawn to the Jefferson Memorial and the tall shaft of the Washington Monument." Early that final morning in the White House, "in my robe, with a cup of coffee in my hand, I made a last pilgrimage into all the rooms on the second floor. This was partly the housewifely need . . . but mostly just to stand still and absorb the feeling of the Yellow Room and the little Lincoln Sitting Room." Diplomatically, she had the staff remove the portraits of herself and Lyndon from the East Hall.

Finishing her sentimental look at their home for more than five life-altering years, she went down to the state rooms and found "the floor was alive with butlers and cleaning people." The last traces of the Johnson era were being swiftly erased. Outside, crossing the North Portico for the last time, ready to leave for the Capitol, "I looked up and there, glued to one of the windows, were the faces of John Fickland and Jerman and another butler. . . . And on the steps smiling and blowing kisses were many members of our staff."

Good-byes can be wrenching for the staff who have shared the

White House and the lives of the family for years, who have gone through ups and downs with them—and are anxious about the new family. And they can be strained: Grace Coolidge virtually had to force her husband to bid farewell to the staff who had served them for more than five years; the Hoovers said no good-byes at all (though they took seven of the White House staff for their estate in California).

Back at the familiar LBJ ranch, Lady Bird retired early "with a line of poetry reeling through my mind. I think it's from India's Love Lyrics. 'I seek, to celebrate my glad release, the Tents of Silence and the Camp of Peace.' And yet it's not quite the right exit line for me because I have loved almost every day of these five years."

In 1953, when it came time for the Trumans to leave the White House, Bess had no desire to linger. She was eager to leave the spotlight, eager to shed the burden, eager to go home to Independence, to another big white house, the one her grandfather had built in the 1860s, a white frame showplace fancied up with gingerbread and stained-glass windows, the house where Harry had returned a cake plate as an excuse to renew a friendship with the blue-eyed girl he had loved when they were six-year-olds in Sunday school, the house where their wedding reception was held.

Even powerful columnist Drew Pearson, a merciless Truman enemy, praised Bess: "If not for her quiet, diplomatic expertise and regal bearing, her husband would be in a lot more hot water than he's presently in." Pushed against her will into the role of celebrity, Bess never felt right for the part; she was just Bess Truman—which was her special appeal.

And Harry Truman? How did he leave his turbulent presidency? Without regrets, neither for the personal sparring with his critics nor for any of the controversial decisions he made. His last act before leaving the White House was meticulous—he returned a pen he had borrowed from an aide.

Following the transfer of power, the Trumans went to Secretary of State Dean Acheson's home for a farewell luncheon with the cabinet and their wives. Georgetown's P Street, Truman recalled with pleasure, "was full of people who cheered as if I were coming in instead of going out." When he and Bess headed back to Missouri, traveling one last time aboard the presidential railroad car, the *Ferdinand Magellan,* thousands jammed Union Station, shouting, cheering, reaching out to him. That day, he said proudly, the crowds were honoring not the President but "plain Mr. Truman now, a private citizen."

As the Trumans said good-bye to the staff, who had been particularly devoted to them, Chief Usher Howell Crim mused, "I watched the Hoovers and the Roosevelts grow into something completely different from what they were when they moved in. They left as different people. But the Trumans—Margaret grew up, that was all."

An alpha President who has been a success in the job will probably feel regret as he packs up to leave the White House; there's never enough time to accomplish all he had hoped to do. Theodore Roosevelt, as his eighth year in the White House drew to a close, came to rue the day he had too quickly ruled out a third term—he had loved being President and reveled in White House life. To his closest aide, Major Archie Butt, he confided—with refreshing honesty—"When you see me quoted in the press as welcoming the rest I will have . . . take no stock in it, for I will confess to you confidentially that I like the job. The burdens of this great nation . . . will be laid aside with a great deal of regret, for I have enjoyed every moment of this so-called arduous and exacting task."

Bill Clinton, despite his close call with impeachment, was a happy fellow in the White House and was reluctant to leave it. He loved it. After eight years there he showed no traces of aging except for a full head of graying hair. Being there, being President, everything about it, was the culmination of a lifelong dream that took hold the day he, a schoolboy from Hot Springs, Arkansas, met John

F. Kennedy in the Rose Garden—the genesis of a life in politics. He had come into the White House as governor of one of the bottom-tier states; he was leaving as the preeminent world leader—and if that weren't enough, his wife had made history as the only First Lady to win high office on her own. The impact of the White House on their lives was beyond measure.

After eight years, Clinton, leaving office, still felt the special magic of the place. "Even today," he said, "I feel a sense of awe. I feel a sense of history. I feel I am carrying on a conversation with all my predecessors." He loved the pageantry, the power, and, yes, the problems, the diligent work of trying to solve the issues eating at the nation and endangering the world.

In 2000, at his final evening at the Kennedy Center's annual awards for excellence in the arts, Clinton declared to the packed house: "Let me say this to you: this night, and every night before, has been a profound honor for Hillary and me. You may find people who do this much better in the future; you will never find anybody who loves it as much." In the Clinton years, the performing arts knew that they had a friend in high places—and he had made friends with high-placed artists.

And Hillary? Political pundits like to speculate that perhaps she hasn't said farewell to the White House at all.

AFTER ONLY THREE months in the still-unlivable White House, Abigail Adams hurried back to her beloved Massachusetts and their welcoming home. Contented, she declared, "We retire from public life. If I did not rise with dignity, I can at least fall with ease, which is much more difficult." But in a private letter she was more trenchant about her brief time in Washington: "My residence in this city has not served to endear the world to me. I am sick, sick, and sick of publick life."

A belated grace note softened John Adams's hurried departure to escape Jefferson's ascent to power. Many years later the two old lions, thinking back on the miracle they had wrought, resumed their friendship through a flow of thoughtful letters, and on July 4, 1826, fifty years to the day after the Continental Congress declared independence, both departed the world. As if by design, the still-new United States gave up the two visionaries who, acting together and clashing apart, had done most to conceive and bring life to the noble experiment. "The whole nation," reported a sorrowing Perley Poore that day, "clothed itself in mourning."

Twenty-eight years after Abigail and John returned to their beloved Massachusetts, their son, John Quincy, left the White House on an equally sour note. His one term as president, he would say, brought "my most miserable years." Two years later he added a historic postscript to his long career, returning to Washington as a member of the House of Representatives, where no other former chief executive has ever served. "My election as President of the United States," he declared, "was not half so gratifying to my inmost soul." For seventeen years he was the revered elder statesman and antislavery champion. In February 1848, he had just cast a vote when he suffered a stroke; after two days in a coma he died in the Speaker's Room, a few yards from his desk. From the time he was the precocious fifteen-year-old secretary to the American minister in Moscow until the day he died, John Quincy Adams had been at the heart of government, but only in his last calling did he capture the people's affection.

IT WOULD BE understandable if Andrew Johnson, in 1869, had turned his back on the Executive Mansion, embittered by four years of hostility from the Congress and his own vindictive cabinet. The Senate rejected—by one vote—the political move to impeach him; the outcome had seemed so certain that the Senate leader in line to succeed Johnson as President was already making plans to move into the White House.

To the contrary, a jubilant throng gathered at the White House to celebrate his victory over the Radical Republicans, and the White

House once again came alive. Washington society writers, assured that Johnson would remain in the White House, tossed verbal bouquets: daughter Martha, his hostess, became "ravishing," her entertaining "brilliant." Johnson's farewell reception was the biggest ever. "A vast concourse assembled to pay their respects to the retiring Chief Magistrate," cooed social columnist Laura Holloway. "Some five thousand sought admittance in vain, while fully as many must have gained entrance." Inside, they "vied with each other in expressing admiration for the honest, upright conduct of the retiring Executive and his charming daughters." Consider the difference made by a single senator's vote.

Five years later the legislature of Tennessee sent Johnson (after two failed bids) back to the Senate—he had previously served a term there in his forty-year career in politics at every level, from alderman to President. Like John Quincy Adams, Johnson declared that Tennessee's reaffirmation meant more to him than his time as President. Only five months after he took his seat, among the senators who had voted to impeach him, he was felled by a stroke—again like Adams. Death deprived him of the chance to restore his reputation, unblemished until his traumatic years in the White House, where he had been the wrong man at the wrong time.

DEFEAT INFLICTS ITS sting in different ways on different psyches. Barbara Bush, resilient, feisty, and topping the popularity polls, found it hard to accept that Bill Clinton, with his tarnished reputation as a womanizer in his years as governor of Arkansas, had defeated her husband by a solid margin. Sixty years earlier, in 1932, Lou Hoover, confident of the outcome of the election when she boarded a train to vote in California, returned to the White House crushed, unable to face the finality of packing. Extra domestic help had to be called in at the last moment to help the White House staff get the Hoovers out of the south door as the Roosevelts were on the north doorstep.

Losing is brutal for a proud President, and rejection is often even harder for his wife, who hurts for both of them. Jimmy Carter had to dig deep into his resolute faith when he lost to Ronald Reagan; Rosalynn, devastated, slipped into depression. "It took me a while," she acknowledges. "I was bitter. I thought I was coming home from the White House and would be bored to death for the rest of my life."

The President's mother, Miss Lillian, wrote of her very different reaction to the defeat: "I said, 'Good!' It wasn't a blow to me—I wanted him out. My whole family had been split wide open from Jimmy being President."

Returning to all that was familiar in Plains, Jimmy and Rosalynn recovered their spirit and spread their wings to engage in humanitarian work around the world; their Carter Center programs attack devastating diseases and nurture democracy in some fifty countries. They spend about one third of their time traveling the globe, and in this country Rosalynn continues to be honored for focusing public attention on mental-health issues. Not resting on the laurels of his Nobel Peace Prize, the former President is an international advocate of human rights and monitor for troubled elections; he has worked with hammer and nails for Habitat for Humanity and has written a score of books, ranging from his memoirs to poetry, children's stories (illustrated by daughter Amy), and most recently a novel. Somehow he found time to build a walnut coffee table that brought $275,000 at an auction benefiting the Carter Center. The Carters definitely are not bored.

Unlike most defeated Presidents, William Howard Taft bore no resentment over leaving the White House. As his term came to an end, he told a friend, "The nearer I get to the inauguration of my successor, the greater relief I feel." Journalist David Barry, who covered Taft and eight other Presidents, was certain that "Mr. Taft never truly wanted to be President. He always knew himself to be a poor politician." In his view Taft had been pushed into running by "the members of his family and the politicians." Taft's ambitious wife, most of all, had set the presidency as the goal.

It was just as well that he was diffident, for he carried only two states, Vermont and Utah, in his bid for reelection in 1912. In a brass-knuckles chapter of politics, Taft had been scuttled by his onetime mentor, Theodore Roosevelt. TR, who could have won a third term easily in 1908, stuck with the unwise declaration he had made after first being elected, ruling out a third term. Then his ill-conceived third-party bid four years later, on the Progressive Party ticket, effectively won the election for Democrat Woodrow Wilson, ending sixteen years of Republican control and changing the course of early-twentieth-century America, not least of which was America's entry into the First World War.

Before leaving town after the inauguration, Taft attended—and heartily sampled—the new President's inaugural luncheon in the State Dining Room while his disheartened Nellie remained upstairs, with the excuse of last-minute packing. In her memoirs she philosophized, "There is always bound to be a sadness about the end of an administration, no matter how voluntarily the retiring President may leave office, no matter how welcome the new President and his family may be." Not that her retirement was voluntary, or the new family welcome. She did not say good-bye to the staff, not even to her personal maid, one of the few who was truly fond of her.

But everything came up roses for the former President in political exile: after several years on the Yale Law School faculty, Taft was named chief justice of the United States Supreme Court by President Harding, the position he had always yearned for. There he happily spent the remaining nine years of his life. In politics timing is everything.

Like Taft, James Buchanan, who had done nothing to address the widening breach between North and South, was eager to leave the White House, relieved to turn over heavy burdens and lesser joys to Abraham Lincoln. "I am heartily tired of my position as President," he wrote to the widow of President Polk. "I shall leave . . . with much greater satisfaction than when entering on the duties of the office." At the Executive Mansion he greeted Lincoln warmly: "If you are as happy, my dear sir, on entering this house as I am on leaving it and returning home, you are the happiest man on earth." The nation would share the happiness of his departure—Lincoln, a President with strength lacking in Buchanan, would face the problems and save the Union.

Twenty years earlier, President Van Buren had reacted to his defeat by William Henry Harrison in the same vein, declaring: "As to the presidency, the two happiest days of my life were those of my entering upon the office and my surrender of it."

SOMETIMES A PRESIDENT has departed with feelings beyond sadness; Millard Fillmore was bitter about leaving the Executive Mansion in 1853. After filling the President's post for nearly three years following the death of Zachary Taylor, he was urged to run for election for a full term but after fifty-three ballots in a bruising convention, his Whig Party rejected him in favor of General Winfield Scott, who then lost to Democrat Franklin Pierce.

Fillmore's resentment soon turned to devastation—at the inauguration, a blustery day with slushy snow underfoot, his deeply loved Abigail caught a cold, which overnight worsened into pneumonia. For three weeks he and their children remained at her side in their Washington hotel suite, but despite all efforts she died without ever returning to their home in Buffalo. The thirteenth President left Washington heartsick, rejected, and filled with rancor toward his party. Four years later, however, he attempted a return to power under the banner of the American Party ticket, the Know-Nothings, a despicable group based on hatred of immigrants and Catholics. With his running mate, Andrew Jackson's nephew Andrew Jackson Donelson, Fillmore came in a lagging third.

ON MARCH 4, 1889, Grover Cleveland and his wildly popular young First Lady were saying good-bye to the White House staff. Despite her husband's defeat for reelection, Frances was not downhearted. "Now, Jerry," she said to a houseman, "I want you to take good care of all the furniture. . . . I want to find everything just as it is now when we come back. For we are coming back four years from today." Incredibly, she was right in her outlandish prediction. During their four years of political exile in New York, Frances had become a confident wife and social leader, eager to reclaim her position as First Lady in 1893.

Following their second term, the Clevelands retired to Princeton. The former President relished the role of elder statesman and was a frequent contributor to *The Saturday Evening Post*, on such diverse topics as politics and "A Defense of Fishermen." The former First Lady was described by a smitten reporter as "a sort of patron saint and goddess in human form."

Early in 1913, five years after her husband's death, Frances returned to the White House as guest of honor at a nostalgic dinner given by Nellie Taft. The fifty-two guests included Cleveland cabinet members or their widows and Esther Cleveland, the White House baby now a grown woman. And there was another special guest: Princeton professor Thomas J. Preston, whom "Frank"—still only forty-eight years old—would marry the following month. She was the only former First Lady to remarry and give up the presidential name until Jackie Kennedy shocked the nation with her marriage to Greek shipping tycoon Aristotle Onassis. The two, not insignificantly, were the most glamorous young First Ladies ever to grace the White House.

NO PRESIDENT'S LEAVE-taking was as pathetic as that of Woodrow Wilson, whose life in the White House was bookended by tragedy— his first wife's death at the beginning, his stroke at the end. Yet the years between were suffused with a new and passionate love.

The White House had been all but locked up in his last year

while his wife functioned as virtual President. Yet—impaired by a stroke and crushed by the defeat of his prodigious effort for the United States to lead a League of Nations—Wilson, against all rational thought, clung to his hope for a third term. His Secret Service agent, Colonel Edmund Starling, later related the President's scheme: at just the right moment an aide would have Wilson's photograph thrown on the screen at the Democratic convention, which would stampede the delegates into nominating him. The President's orders were carried out, but—to no surprise—the fatuous plan failed. It was the tragic attempt of a man who had been robbed of his link to reality and could think only of retrieving the power that had somehow vanished. Such is the siren call of the White House.

GEORGE WASHINGTON, WHO could have stayed in office indefinitely, decided that two terms were enough, setting a precedent that was never broken until Franklin Roosevelt went for a third, and then a fourth. After that, Congress imposed a limit of two, surely a wise restraint. While most Presidents have been relieved to be free of the millstone after eight years—or even four—others have wished for a third term.

President Grant—at the urging of First Lady Julia—sought a third term, but the cloud of corruption hanging over his administration led the GOP to turn him down. Grant had been obliged to apologize formally to the Congress for his "errors of judgment," and in his final address to Congress he offered the lame excuse that "it was my fortune or misfortune to be called to the office of Chief Executive without any previous political training." Commented journalist Perley Poore, "It was as if his glory as General would more than atone for his deficiencies as President." Grant, while honest, Poore observed, "appeared unable to discern dishonesty in others."

Julia was heartbroken at leaving the White House after eight glorious years. "Oh, Ulys," she wept, "I feel like a waif, like a waif on the world's wide common." Four years later, after a two-year grand tour of the world, fêted by the crowned heads and potentates who had been their White House guests, Grant—and Julia—again tried for a third term. After thirty-five ballots the deadlocked convention turned to James Garfield, who had, perhaps not coincidentally, led the opposition to Grant.

After leaving the White House, where they had grown accustomed to the best, Grant, in the get-rich bull market of the early 1880s, entrusted his holdings to his son "Buck," ostensibly a financial expert. Buck managed to lose it all in a swindle that sent his partner to prison. The former President, in debt and struggling against throat cancer, turned to writing his memoirs as his only way to provide for Julia. His life story, published by his old friend Mark Twain, was more than a best seller—it reaped nearly $450,000 in royalties, a huge amount in those days, and is still regarded as the finest presidential autobiography. Four days after completing the manuscript, the old soldier died. Julia lived stylishly on *Personal Memoirs of U.S. Grant* for seventeen years and, thinking her own recollections would meet a similar reception, wrote a flowery little memoir, asking such an exorbitant price that no publisher, not even her vastly rich friend Andrew Carnegie, would agree to take it on. It was finally published by Southern Illinois University in 1975 as one First Lady's life story—with long correction notes appended.

GRANT'S SUCCESSOR, RUTHERFORD Hayes, left the White House amid catcalls, after a mere single term. *The Washington Post* thundered, "There should be no spot of ground on the continent to give him harborage or shelter save the few feet of earth needed for a nameless grave. Exit Hayes the fraud. Eternal hatred to his memory." The *Post*'s screed was likely fueled by Hayes's conciliatory policy toward the South, infuriating Old Guard Republicans; that he revived Civil Service reform, breaking up the treasured spoils system, added to the editor's choler. Hayes declared himself quite pleased with his presidency. To an old classmate at Kenyon College he wrote, "Nobody ever left the Presidency with less regrets, less disappointment, fewer heartburnings, nor more general content with the result of his term (in his own heart, I mean) than I do."

But it was not to be a happy leave-taking. Following the inauguration of his successor, James Garfield, the Hayeses boarded a special train that would take them back to Ohio. Before reaching Baltimore, there was a terrible jolt and screeching of metal—somehow the train had crashed into another on the same track. Hayes was hurled from his seat, unhurt; in other cars two were killed and a score seriously injured. It was a traumatic end to their pleasant term in the White House; however, in his twelve remaining years, he and Lucy lived a contented life centered on philanthropic causes, quite pleased to have the memories and none of the burdens of the White House.

THE FIRST VICE President to reach the White House by political happenstance, John Tyler, was rebuffed by all sides when he sought election in his own name. Still, he was content to return to his Virginia estate with his charming new wife, Julia, who in less than a year had cut a swath through Washington.

His departure revealed to him the evanescence of an ex-president's power. He arrived at the wharf with his wife and a jumble of children, servants, and mounds of luggage just as the steamer sounded its whistle and was moving out into the Potomac. As Perley Poore described the scene, "Someone sang out, 'Hello! Captain, hold on there! Ex-President Tyler is coming. Hold on!'" The captain, an old Whig—Tyler's party before an acrimonious split—"pulled his engine bell violently and shouted, 'Ex-President Tyler be dashed! Let him stay!'" With that, he swung his vessel out into the Potomac, leaving yesterday's First Family stranded on the dock.

In retirement Tyler was obviously a happy man as his young wife bore him a new family of seven children, bringing his total offspring to fifteen. Their lives, from first birth to last death, stretched across thirty presidencies and an amazing 132 years.

SOME CHANGES REPRESENT more than a replacement of one set of White House residents with another; they signal the beginning of a new era. Tyler's departure in 1845 marked the end of Virginia's dominance of the presidency. Four of the first five were elected from the Old Dominion, and though William Henry Harrison, the ninth President, whom Tyler succeeded, was elected as a westerner, an Indian fighter from Ohio, he had been born and brought up on his family's fine Virginia plantation, a far cry from the "log cabin and hard cider" slogan of his campaign.

But Old Tippecanoe represented the changing nation as its power and influence shifted west, while Tyler, his much younger Vice President, was a remnant of postcolonial aristocracy. The social whirl of Julia Tyler's short tenure, Perley Poore commented, marked "the end of the Cavalier reign within the White House, which was soon ruled with Puritan austerity by Mrs. Polk." At regular intervals there have been such overnight changes of impact, of eras— Buchanan to Lincoln, Garfield to Arthur, McKinley to Theodore Roosevelt, Hoover to Franklin D. Roosevelt, Eisenhower to Kennedy—whose White House families demonstrated a sharp turn to the modern.

For the Eisenhowers, who were among those who made the White House very much their home, leaving was sweet sorrow. The world's great general and his lady were accustomed to being catered to, accustomed to having legions of staff snap to when they spoke. And, yes, accustomed to being celebrities. The White House suited them.

In a sentimental close to the public chapter of a celebrated life, all members of Eisenhower's official family and a few close outside friends came for a farewell evening, beginning with drinks in the family quarters—which few of them had ever seen—followed by a roast-duckling dinner in the State Dining Room, reminiscing and dancing in the East Room. The Marine Band in their scarlet tunics, the Army Chorus, and the Army Singing Strings played and sang their hearts out for their old commander. And when the evening ended with "Auld Lang Syne" and "Bless This House," there was a lump in the throat as old friends and colleagues recognized the moment as the end of a special time they had shared with one another and the country.

Ike's grandson and Nixon's son-in-law, Dwight David Eisenhower II, now a historian and writer, speaks with a unique perspective of two very different presidents, two very different White Houses. In his dual connection, David saw the mansion smoothly adapt to its residents. Contrary to their public images, David observes, Eisenhower was much more formal than Nixon, who was quite informal: "Ike liked hierarchy, protocol, place, boundaries." Eisenhower was shaped by the military; Nixon, to some extent, was shaped by California.

Their wives, Mamie and Pat, were the products of equally different backgrounds. Mamie was the spoiled daughter of a large, lively family of very considerable wealth; Pat was born into a miner's family always scrimping to get along and was thrust into responsibilities beyond her age when her mother died. Bubbling Mamie was a national favorite; Pat, who did more for the White House, was never given the credit she deserved.

In a thoughtful interview on C-Span, David offered a significant insight into his grandparents' marriage: "They were a war couple. It was a huge dimension in his life, but my grandmother was cut out altogether. He became a world figure, but that rise was not really shared by her. She had been displaced; she didn't understand her husband anymore." But in the White House Mamie was fulfilled— she was a star in her own right and a valued partner to her Ike.

The Eisenhower White House was the well-run home of one of the most popular figures ever to live there and it was also a second home for David and his three sisters, who had "the run of the place." But their grandfather was firm that, unlike the typical nineteenth-century adult sons, who so often worked for their fathers and lived in the White House, his family would maintain its own residence, a modest house across the river in Virginia, and his four children would have normal lives. John (now the oldest living White House child) was assistant director of the White House military staff; later he would find his real calling as editor of his father's extensive papers and as a distinguished military historian.

In his second White House experience, David saw the Executive Mansion become a redoubt protecting the embattled Nixon family against the enveloping tide of Watergate, which ultimately swept the President into history. During that period Washington journalists noted that David, then a law student at George Washington University, maintained a careful distance from his wife's public defense of her father. Looking back, he described his feelings in his two leave-takings: "Regret in 1961, relief in 1974—we felt that everybody would move on, things would be better." He was right: he and Julie have lived a full life as writers, speakers, and parents of three daughters.

TWO DAYS BEFORE he left the White House, Thomas Jefferson penned a letter to his old friend Pierre du Pont de Nemours: "Never did a prisoner, released from his chains, feel such relief as I shall on shaking off the shackles of power." After eight years as the first full-fledged White House resident, Jefferson was leaving a lifetime of public service that was interwoven with the nation's history.

The third President, whose vision was reflected in every aspect of the Executive Mansion and the Federal City, chose not to ride to the Capitol in the coach with James Madison, his friend and handpicked successor. Exercising the same democratic impulse he had shown when he walked to his own inauguration, he followed on horseback behind the new President, in the company of other riders. Once again, he would be Thomas Jefferson, citizen. He had set the pattern, launched the President's House, resisted the lure of royal pomp, encouraged reading, music, agriculture, invention; he had vastly enlarged the country and envisaged an even greater reach when he dispatched Lewis and Clark to discover what lay to the west.

He departed the President's House as simply as he had arrived. Mounting his horse, he rode 140 miles toward Virginia's Blue Ridge Mountains, back to his true heart—his family, his books, his gardens, his Monticello.

ONE HUNDRED AND sixty-eight years later, another President who had brought a new day to the White House, Gerald Ford and his bold Betty also headed west after the inauguration of his successor. Boarding a presidential helicopter on the Capitol grounds, the Fords circled over monumental Washington for a nostalgic farewell to the city where they had spent their life together and the white-columned mansion where fate had taken them. Then a presidential plane (not *Air Force One*—power shifts in an instant) headed west, returning them to the real world of their Colorado retreat.

Jefferson left the White House relieved; John Adams left dejected; Lyndon Johnson left broken by war; Theodore Roosevelt left regretting that he had promised not to run again; Nellie Taft so wished to stay on; Rosalynn Carter was heartsick; Louisa Adams was happy to escape; Julia Grant could not face leaving her golden life. The children who had not already taken wing were sometimes glad, sometimes sorry, often both. Their thoughts as they departed were as different as the individuals had been, but without exception all left knowing that the White House had indelibly marked their lives.

fourteen

Footprints on History

———⊰◦⊱———

No matter if she's there for only eight months, like Julia Tyler, or twelve years, like Eleanor Roosevelt, whether she comes with the experience of a governor's mansion or a Vice President's ceremonial world, the White House changes the President's wife. All First Ladies, even those who came as unnoticed wives and were eager to return to that status, have been transformed into something larger than themselves, into women whose influence continues long after they leave the stage of the Executive Mansion, whose aura will never vanish.

Today, her public role begins when she crosses the North Portico on Inauguration Day to enter the Executive Mansion that will be hers. There is no place for a First Lady who resists. A contemporary wife understands that she is being given the opportunity to make a difference, to do something in her own way for the country. She now has influence—power—through what Hillary Clinton calls the "white glove podium" and the "derivative power" stemming from wifehood. Nancy Reagan lightly commented, "For eight years I was sleeping with the President, and if that doesn't give you special access, I don't know what does!" In Washington access is power; Nancy had it and used it.

The First Lady is the most effective advocate in chief for the causes that are special to her. In Laura Bush's view, "Americans want the First Lady to do whatever it is she wants to do, and are supportive no matter what she might want to work on." The operative words there are *do* and *work;* presiding over requisite social events and shaking many hands, as First Ladies through Bess Truman and Mamie Eisenhower did with differing enthusiasm, is no longer enough.

A reluctant Lady Bird Johnson was at her husband's side when he agreed halfheartedly to run as Vice President, and was horrified, along with the world, that tragedy thrust him into the presidency. In her new role "she recognized from the first moment the enormity of the responsibility," her daughter Luci says, speaking for her mother, whose speech—but not her mind—has been impaired by a stroke. With her lifelong love of nature, Lady Bird focused on the environment. "There was a sense of urgency," says Luci. "We were destroying our American landscape—and you can't get it back. Natural beauty, Lady Bird would say, 'feeds the soul and lifts the heart.'"

Lady Bird took her White House podium further afield than any other First Lady. In her five years she traveled two hundred thousand miles spreading her gospel of protecting the environment and the national parks, exhorting Americans to add grace to their towns, beautify their roadsides, and preserve every remnant of their local history. And to promote her husband's Great Society programs, she visited, with singular empathy, urban ghettoes and North Carolina's remote mountains "to draw back the curtain, to awaken and alert the country" to entrenched poverty amidst prosperity. "I like to get out and see the people behind the statistics," she said, and whatever Lady Bird reported to the President, he believed. In the range of her concerns, observed her friend and champion Laurance Rockefeller, "She's a role model for leadership responsibility for women. That's a big part of her legacy above and beyond the environment."

With Lyndon Johnson as her lobbyist, Congress passed the Highway Beautification Act banning billboards and junkyards from interstate highways. (At a Christmas reception for the Congress, the President stood on a chair and declared—only half joking—that he would keep them in session "until you pass Lady Bird's bill.")

The Lady Bird expeditions won media coverage like no other First Lady's, thanks in equal measure to her deep commitment, her poetic turn of phrase, and to a chief of staff who would have made P. T. Barnum envious. Liz Carpenter, a former Texas journalist, used all of America as a prop for the First Lady. From the craggy coast of Maine to the awesome redwoods of Northern California (where a grove is named in her honor), Lady Bird's press corps happily traipsed after her by rubber raft, ski lift, horse-drawn surrey, Mississippi paddle wheel, and Texas orchard wagon—with results spread across television and newspapers. (Occasionally the glowing coverage sparked jealousy in her husband; on an important trip to rally his Vietnam allies in Asia, LBJ groused to Liz Carpenter, "How come Lady Bird gets all those good stories on the front page and there's nothing but bad stories about me?")

Her impact is still recognized—and grows—all across the country, not just for what she accomplished in her White House years but for the movement she did much to launch. Washington still blooms with Lady Bird's dedication; beautification (the term fell far short of her concerns, but no one came up with a better one) is a civic enthusiasm all across the country. On the banks of the Potomac, Lady Bird Johnson Park blooms, and in her home city of Austin the Lady Bird Johnson Wildflower Center advances the science of natural landscaping. With her White House crusade for a more beautiful America, Lady Bird has given the country the most lasting legacy of all First Ladies. The nation agreed. At a joint session of Congress in April 1988, she was awarded the Congressional Gold Medal for her "impressive career in her own right."

MAYBE YOU ARE driving along Barbara Bush Avenue, passing by Barbara Bush Children's Hospital or one of six Barbara Bush schools. You pop into a bookstore to pick up the newest Barbara Bush best seller, or perhaps you've just visited the Houston Livestock

Show and watched a fine heifer named Barbara Bush place eighth in the judging. You wouldn't have to ask who this eponymous Barbara Bush might be—you'd know. Though Barbara Bush was First Lady for only one term, in those four years she won admiration and influence that stayed green through two succeeding Democratic terms and was refreshed when her son secured for her a place in history alongside Abigail Adams, the only two women to be both First Lady and First Mother. (And Abigail, sadly, did not live to see her son as President.)

When her books come out, the former First Lady is fought over by the talk shows. She is more relaxed in interviews than the pros asking the questions; her easy humor is self-deprecating, her timing for laugh lines perfectly honed. The First Mother lets the audience in on her President-son's practical jokes and says without hesitation that she gives him advice. Pause. "He never takes it." (When her husband was President, she also weighed in with her views: "Show me a wife who doesn't offer advice and I'll show you one who doesn't care very much," she commented in her memoirs.) She is called "the Enforcer" by her family because she tells everybody what to do. (In one interview she demurred at the title, leading her husband to interject, with just a trace of emphasis, "But you do, Barbara, you do!") Not long before the Republican convention of 2004, Jay Leno asked Laura Bush about her mother-in-law's nickname. "She was pretty intimidating," Laura acknowledged with a chuckle. "She still is sometimes!"

Barbara's legacy has greater substance than her books about her golden life in and out of the White House. As First Lady she reinvigorated the White House Historical Association that generates large sums of money for the Mansion, and she continues to lead the literacy crusade she launched with her leverage as First Lady, encouraging children—and parents—to read. Through her speeches all around the country and high-stakes fund-raisers, she has raised millions of dollars for the Barbara Bush Foundation for Family Literacy.

It is a happy coincidence that her daughter-in-law brought to the White House a similar concern for reading, especially to children. She did not adopt that cause to please the Enforcer—long before she met the Bushes, Laura Welch was a teacher and librarian. Expanding on her innovation as First Lady of Texas, she organized the National Book Festival involving the White House and the Library of Congress, offering the unmatched prestige that attracts major authors.

Laura came into the White House asserting that she would not be a public figure, probably to emphasize the difference between herself and Hillary Clinton. Instead, she has become a major player, comfortable in any setting, a woman at ease in her own skin. She realized, particularly after the attacks of September 11, that there was much she could do to serve the country and her husband's agenda.

Across America she sounds the alarm about heart disease in women. Her boldest mission came in March 2005 to still–dangerous Afghanistan—keeping her departure plans secret from the media and, she said, even from her husband. She was a smiling touch of home for U.S. troops, but her deeper purpose was to bolster education for girls and basic rights for women. Two months later she took her message into the volatile Middle East, addressing the World Economic Forum in Jordan with visits to Israel and Egypt. All through the visit gunships flew with her helicopter and sharpshooters stood watch at every stop.

In the 2004 campaign Laura was everywhere, not as an adjunct to the President but as a confident speaker, the "big draw" and the very big fund-raiser who promoted her husband and stayed "on message." Had you tuned in late, you might have found her bantering with Jay Leno on *The Tonight Show*, sometimes topping his laughs. Enjoying the freedom of a second term, she nudged her husband aside at the power-studded White House Correspondents' dinner and wowed them with a speech kidding him—territory no other First Lady has ever dared to enter. Some even saw her imprint on the 2005 federal budget: as program after program was cut, it was noted that the Agency for Library Services, dear to her heart, received a nine-million-dollar increase.

"Do I aspire to be up there at the top of the list?" she asks rhetorically. "Absolutely. While you have this forum, you want to be as constructive for your country as you possibly can." Laura discovered what a long line of First Ladies before her came to understand: the White House is a force; you do not leave as the same woman who arrived, uncertain, anxious, and even reluctant, on Inauguration Day.

Columnist Helen Thomas, dean of White House correspondents, attests to that. Over the years, she has kept her sharp reporter's eye trained on nine First Ladies and has observed a pattern that is repeated in each of them. "In the White House they get their own identity," she explains. "The First Lady becomes a person in her own right. She is no longer Mrs. Ronald Reagan or Mrs. George Bush—she becomes a separate entity as Nancy Reagan, Laura Bush. They suddenly realize what they should do for their country. At some point every First Lady develops a social cause. She realizes that she can wave her magic First Lady's wand and the world will beat a path to her door."

Laura Bush's commitment to libraries as the heart of learning took on unexpected urgency in the wake of Hurricane Katrina, which devastated the libraries of two hundred schools across the Gulf Coast. Staggered by the loss of books, materials, equipment—and hope—the First Lady led her Laura Bush Foundation into action, soliciting private corporations and the generous public for donations to restock the libraries. By the first anniversary of the catastrophe, she had collected $1.5 million, and the total continues to grow.

The singular influence that is available to a First Lady enabled Laura to supply not only books but a measure of happiness to storm-tossed Gulf Coast children.

HELEN "NELLIE" TAFT is little remembered and not much liked when she is, yet in one term she left two legacies that became cherished traditions. In her first hour as First Lady, in 1909, she showed that she had a mind of her own. When Theodore Roosevelt, the outgoing President, chose to depart directly after the inaugural ceremonies, Nellie announced that *she* would accompany her husband on his triumphant drive to the White House.

In her purple satin suit and lampshade hat spiked by an egret plume as audacious as a personal banner, she brushed aside the protests of the Congressional Committee on Arrangements and planted herself in the carriage beside her husband, who beamed his approval. In her *Recollections of Full Years,* Nellie admitted to "a little secret elation in thinking I was doing something which no woman had ever done before." Daring to signal that this President's wife was a partner in his achievements, Nellie Taft set a precedent that has been followed by every successor, now an official recognition of the role the public expects the First Lady to play in the White House.

Assuming her power as First Lady, Nellie set in motion a dream that has enchanted many millions of Washington visitors for nearly a hundred years: the Japanese cherry trees that each April garland the Tidal Basin and most of monumental Washington, masses of pink froth that take the breath away. Nellie had visited Tokyo while her husband was governor-general of the Philippines and had never forgotten its cherry blossoms. She conceived her grand plan for the capital, and when all of the nurseries in America could provide a mere eighty trees, the mayor of Tokyo, much flattered, sent two thousand as a gift of friendship. Lamentably, the first shipment was insect- and fungus-infested, whereupon the mayor sent two thousand replacements.

On March 12, 1912, the First Lady and the Japanese ambassador's wife turned the first spades of earth, and so began Washington's world-famous spring fairyland. While Lady Bird Johnson is praised anew every year, there is little mention of the First Lady whose vision and drive first beautified the nation's capital. The rather haughty, rarely smiling Helen Taft was never popular, but she left a lasting footprint—and a little-known footnote as the first wife of a President to drive herself around town in her own little electric car.

Yet some First Ladies leave virtually no trace. One such was Florence Harding, who was neither charming nor beloved. Marietta Minnigerode Andrews, an artist who for thirty-two years was an active participant in the Washington social scene, contributed an acerbic summary of the Hardings: "I think there have never been a

President and a President's wife who have made so transient an impression upon the life of the capital as Mr. and Mrs. Warren Gamaliel Harding." Nor has history treated them any more kindly.

Who today remembers early nineteenth-century First Ladies, other than Dolley Madison, a dominant figure in Washington over fifty years, who was sui generis? Who today is even aware of serious-minded Sarah Childress Polk, the First Lady who first stepped beyond the borders of what was expected of the wives of important men? In the mid-1840s Sarah, who had no children, disregarded precedent to work alongside her husband as his confidential secretary and adviser. Perhaps that bold step was the result of an earlier bold step: she was the first First Lady to have left home to attend a formal academy.

On horseback Sarah, no more than fifteen, traveled with her older brother, a manservant, and her younger sister from East Tennessee to North Carolina, an arduous journey of well over three hundred miles across the Smoky Mountains, to attend Moravian Female Academy in Salem (now Salem Academy and College in Winston-Salem). She studied history, geography, and arithmetic along with the more conventional young ladies' tutoring in needlework and music. For that her father paid twenty dollars a quarter, laundry extra. That early education fed a serious mind that would go well beyond the usual interests of her contemporaries.

Even in her own time, admiration for Sarah was such that throughout her forty-two years of widowhood, everyone of distinction called on her in Nashville, including both Confederate General Beauregard and Union General Sherman (traveling separately). The Union general whose forces held Nashville ordered his troops to respect the distinguished former First Lady.

AS THE WHITE House leaves its imprint on the First Ladies, many First Ladies, in turn, leave their imprint on the White House. In 1889 Caroline Harrison, a skilled china-painter, saw history in the remnants of old presidential china she discovered, dusty and unused in a storage room. With that, she started the invaluable collection that was continued by Edith Roosevelt and ultimately completed by Mamie Eisenhower. Edith, one of the First Ladies most interested in preserving the mansion as a museum as well as a home, created the gallery of First Ladies' portraits, recognizing them as the Presidents' partners in history.

Abigail Powers Fillmore was appalled, when her husband entered the White House upon Zachary Taylor's death in 1850, to find that the Executive Mansion possessed not one book, not even a Bible. As a former teacher (in defiance of her times she had taught school after her marriage), she persuaded Congress to provide two thousand dollars to buy books, and soon volumes were pouring in from such purveyors as Mr. Little and Mr. Brown, whose names are still conjoined in a major publishing house. The President personally kept meticulous records of the purchases.

The White House archives reveal a fascinating list of Abigail's eight hundred choices. Beginning with a Bible, of course, and an atlas, and expanding into a ten-volume set of American biographies, she chose, with professional advice, the works of famous authors, studies of the animal kingdom and human understanding, and reference books on such unexpected subjects as anatomy, astronomy, and civil law. In only sixteen months as First Lady, Abigail Fillmore added an intellectual dimension to the White House.

Lucy Webb Hayes's popularity reflected the times; she was derided by men who took issue with her no-alcohol White House, but she left the White House as "the most idolized woman in America," as one journalist put it. She symbolized the need to deal with a serious problem in the nineteenth century, the prevalence of alcoholism and its damage to the family. It was Lucy's commitment to total abstinence that moved the Woman's Christian Temperance Union to acclaim her as one "who had done such worthy things as to secure to herself a following such as no other member of her sex ever had in this country." (Ultimately, that issue led to Prohibition—which proved ineffective and was repealed in 1933.) Lucy, once again in private life, used her White House cachet in her work for civic causes, making her a Victorian woman ahead of her time.

Lou Henry Hoover, a geologist and Asian scholar, an activist First Lady and early feminist, is little appreciated, overwhelmed by her husband's failed presidency. She did more than lend her name as national president of the Girl Scouts of America, she godmothered the organization from one hundred thousand members to almost a million, and raised the enormous sum of nearly two million dollars for them. She was the first First Lady to make public speeches and nationwide radio broadcasts, encouraging women both to be active volunteers or to follow a career if that was their wish, and most certainly she was the moving spirit behind her husband's order to open up civil service jobs to applicants without regard to sex, and his appointment of more women to upper level posts than ever before. She even encouraged 4-H club boys to help with the housework.

Many years before Jackie set her goal to lift the President's House to museum quality, Lou Hoover traced down four of President Monroe's original French chairs and had them meticulously replicated. It was Lou who converted the West Hall into the comfortable living room enjoyed by every First family since and was first to open the White House to the public at Christmas. At her sudden death in 1944, her grieving husband called her "a symbol of everything wholesome in American life." Yet her contributions are scarcely remembered now, swept away in the tsunami of the Depression.

The footprints left by other First Ladies are as varied as the women themselves. Lady Bird Johnson, the naturalist, had a secluded garden designed as her gift to the mansion; Grace Coolidge crocheted a bedspread, square by square, for the Lincoln bed that remains a White House treasure. Mamie left the insouciance of her bangs, and Hillary showed that the White House can be a stepping stone.

———

DECADES LATER, WHEN American women were beginning to ex-
pand their reach and their role, Pat Nixon, who had been her students' favorite teacher at Whittier (California) High School and a
talented actress in Whittier's amateur theatricals, found herself playing the feminine lead in the Nixon White House. The show had been
in previews for a long time—she had eight years as the popular, supportive wife of Eisenhower's Vice President, the consoling wife when
he lost in the razor-thin race against John F. Kennedy, and again
when he failed in his bid for governor of California in 1962.

But when she finally came into the White House as First Lady, Pat
Nixon was never given the chance to be all that Patricia Ryan could
be. The men around the President—and the President himself?—saw
her only as an appendage to make him look better. As a result she was
unfairly painted as "Plastic Pat," seated on yet another platform, legs
neatly crossed at the ankle, looking admiringly at her husband as he
made the speech she knew by heart. She fulfilled every duty expected
of a First Lady with unfailing amiability, put every nervous guest at
ease. In a permanent contribution to the White House, Pat tapped
new sources to add superb antiques to its state rooms, but she received
little credit—that would always be thought of as Jackie's legacy.

To travel with Pat Nixon on her solo trips was to see her emerge
from the White House cocoon. In Peru, taking comfort and supplies
to victims of a catastrophic earthquake, she landed on a makeshift
runway at the mountain's edge that had even a war-seasoned general
in her entourage gripping with white knuckles. On that mission Pat
played a genuine diplomatic role in healing the rift between the two
nations—but on her return she joined the President at the Grand
Ole Opry in Nashville, only to have him forget to introduce her at
what was planned as her birthday celebration. (Great performer Roy
Acuff smoothly covered the blunder.)

On her own in West Africa Pat built goodwill in three countries.

The cheering, dancing street crowds, the headlines and excited attention, were hers. She charmed a group of Ghana's multiple tribal kings in their exotic robes, endured the inauguration of a president of Liberia in near-heatstroke conditions without even fanning herself, and held what appeared to be lively conversations with the French-speaking permanent president of Ivory Coast. Another First Lady would have been idolized back home, but her husband's staff underestimated her—and her husband himself seemed distant. Pat would always be obscured by the dark cloud of animosity that enveloped Richard Nixon; Pat Ryan's natural spontaneity was lost in the politics of controversy.

Her hard early life had steeled her against what might seem to be any vicissitude—"I am never afraid," she once declared. "I am never tired." But nothing could have steeled her against the public disgrace that would obliterate her husband's achievements.

Among first ladies, Betty Ford is unique, the only one to come into the White House without ever being toughened by a national campaign. A free spirit, she said what she thought. Along with her openness about her breast cancer came her second admirable legacy: some time after moving back to private life, she accepted that she had fallen prey to prescription-drug and alcohol addiction, and she talked about that, too. Speaking out on the topic, she turned her addiction into benefit for others, and in 1982 she was the key force in founding a treatment center for alcohol and drug abuse. Statistics demonstrate the success of the now-famous Betty Ford Center in California. The imprimatur of a former First Lady who was herself a recovering addict lifted the stigma from seeking treatment.

President Ford, turning ninety, reflected on his long, full life and once again praised his wife's courage in publicly discussing her cancer and her alcohol and drug abuse. In less than a full term in the White House Betty made an impact as an independent-minded First Lady and left a legacy that still continues. She maintains that once a First Lady, always a First Lady. "The job never ends—the public never stops seeing you that way," she said on the *Today* show, almost thirty years after leaving the White House. "It's a gift, truly a gift." She treasures that permanent influence: "I'm delighted that I can carry a message. If you can, you should do it—I think that's what we are here on earth for."

At the 1976 Republican convention, which nominated Jerry Ford, Betty told *Time* magazine, only half joking, "After I'm no longer First Lady, I'm going to lobby for a salary for this job. It has long hours and a lot of responsibilities. But I would have it so that a First Lady can't collect unless she works." Years later she mused, "I'm not sure I wasn't right." The role is indeed a full-time job, but Betty's half-serious notion is marred by an unfixable flaw: Who would fire the First Lady if she performed poorly? Edith Wilson's role as de facto President illustrates the flaw. In her study of the Wilsons, historian Phyllis Lee Levin declared Edith "a dissembling and unworthy figure in the history of the American presidency."

Nancy Reagan, a sparkling, soft-spoken Hollywood by-product, has also discovered that a former First Lady can command attention for as long as she chooses. Though her public cause in the White House was her "Just Say No" campaign, an inadequately focused program aimed at fighting teenage drug abuse, her true cause, simply put, was the success of her husband. Watching helplessly as he slipped away, she spoke movingly of her personal tragedy: "Ronnie's long journey has taken him to a distant place where I can't reach him. The saddest part is that we can't share our fifty-two years together."

In 2004 the usually reserved Nancy mounted the white-glove pulpit to embrace a new cause; she became a public advocate for embryonic stem-cell research, the experimental—and controversial—science that may hold promise in the treatment of Alzheimer's, Parkinson's, and other neural disorders. Just as Betty Ford openly calls *Roe* v. *Wade* "a great, great [Supreme Court] decision," an emboldened Nancy stands in opposition to the conservative right and President Bush, who ruled out any expansion in the research she maintains is crucial. She presents the case with scientific knowledge and personal poignancy. "I don't think they understand," she wrote in the AARP journal. "It's not taking a life, it's trying to save countless lives." Nancy is using her White House prestige for a special cause.

Even after President Reagan's death in June 2004, Nancy— suddenly frail and fading—was watching over him, overseeing every detail of his stirring state funeral. And watching over her were her children, who for so long were distant, even estranged, from their parents, now there at her side, the strong right arm for her to lean on. The purpose of her life was now gone. Right up until her only love was laid to rest in the spot he had chosen, and the final California sunset lit the mountain-rimmed sea, Nancy Reagan did it all, as she said, for Ronnie.

The whole world knows Jacqueline Kennedy, who could launch a trend with one photograph, who engineered a newly authentic White House with a Carborundum will cloaked in vulnerability. Guarding against possible less-caring future residents, she created the Fine Arts Commission as a watchdog and the White House Historical Association to raise funds; she persuaded Congress to pass a law placing White House furnishings "of historic or artistic interest" under the direct control of the President or in the care of the Smithsonian. Thanks to Jackie, the mansion's fine pieces will never wind up under an auctioneer's hammer. (What would she have felt when her own prized possessions were sold at auction by her children in 1996, reaping $34 million, and again, in 2005, her lesser household belongings were put on the block by Caroline and brought in $5.5 million. Both sales attracted the ravenous public attention that Jackie hated.)

Jackie's stately courage at her husband's death and her resolute privacy further enhanced her image. (The public willingly erased her impolitic marriage to much older, much richer Greek shipping magnate Aristotle Onassis.) Along with Dolley Madison, Jacqueline Kennedy will be remembered and reconstructed a hundred years after her too-brief White House tenure.

Like Jackie, Eleanor Roosevelt was a product of a privileged world, but unlike Jackie she applied her "derivative power" to the nation's have-nots and can't haves. It is commonly said that Eleanor served as "the eyes and ears of the President," but it seems more exact to suggest that she was directing public attention to deplorable conditions that were her own deepest concerns. Without question, her commitment to the forgotten Americans was deeper, more urgent, than her husband's; she took their case to the President.

Her appointment, by Harry Truman, to the United Nations let her shine her light on human suffering around the world. Even knife-tongued Republican Clare Boothe Luce praised the bête noire of the GOP: "Now the plain fact is, Mrs. Roosevelt has done more good deeds, on a bigger scale, for a longer time, than any woman who ever appeared on our public scene. No woman has ever so comforted the distressed—or distressed the comfortable." In the gallery of First Ladies who left a deep footprint on White House history, Eleanor Roosevelt will always be a dominant figure.

TEN FIRST LADIES, including all of the most recent five, have left a wonderful legacy—their memoirs, sharing their lives in the White House, the good times and bad, told with intimacy and insights lacking in a President's memoirs. While there are a few newsy letters from Abigail Adams in her short three months in the White House and a few from other early wives, Julia Grant was the first to put her story in book form (but couldn't find a publisher), while Hillary Clinton's *Living History* broke sales records in 2003. Lady Bird Johnson, with her fine ear for language, used a tape recorder to capture each day as it happened. "I wanted to share life in this house, in these times," she said. "It was too great a thing to have alone." For a reader, the result is like being there.

Even the less accomplished among them recapture the vast changes in White House life and in the lives of American women over the 132 years separating the wives of the eighteenth and forty-second Presidents. Their books contribute to their permanent footprints on history and will give generations of readers an understanding of the human side of the President's House. Hopefully, Laura Bush whose early career was built around books, will add hers to the collection.

MOST—BUT FAR from all—of the children who have lived in the White House look back on it as a great experience that made an impact on their lives. All would agree with Lynda Johnson Robb that "it made me much more understanding and sympathetic for the people who live there."

Dwight David Eisenhower II, the grandson-namesake for whom the White House was a second home in both his grandfather's and his father-in-law's administrations, traces his career as a historian, author, and academic to his earliest years in the storied mansion. Along with finding its hallowed halls ideal for displaying his collection of baseball cards, "I think even as a five-year-old I understood that the White House was to be revered," he told a C–Span interviewer. "I grew up to be a passionate lover of history—so much was in the air in the White House." It led him, a fellow at the University of Pennsylvania, to the daunting task of writing the definitive biography of President Eisenhower. The first volume won high praise; the project is a work in progress.

Like her husband's, Julie Nixon Eisenhower's books stem from her White House years, especially her well-received biography of her mother. But the most public, and awkward, residue of the White House surfaced when the two sisters, Julie and Tricia, publicly split on how the Richard Nixon Library should be run. The dispute erupted over the bequest of twelve million dollars, or more, from the President's closest friend, Bebe Rebozo, a Florida businessman whose house was encompassed in Nixon's Key Biscayne compound. The daughters' disagreement wound up in the headlines and in a California court. Tricia demanded a board controlled by family and close friends; Julie countered, "Families don't run real libraries, professionals do." The two husbands lined up behind their respective spouses—or perhaps nudged them. Eventually an agreement (pro-professionals) was forged, and the two sisters remain friends.

Though Susan Ford Bales, now a Betty Ford look-alike, was in the White House less than a full term, it gave her a lifetime of experience—the opportunity to study photography with the legendary Ansel Adams; an internship at the *Topeka Journal,* where she learned photojournalism; and her intimate knowledge of the White House provides the setting for her mystery books featuring—yes— a President's daughter who is a photojournalist. "You take that opportunity and run with it—some have, some haven't," she observes. "Or you don't get involved, and stay behind closed doors."

Among her most memorable moments, she says without hesitation, was "the summer of the bicentennial year—1976. Dad and I went up to New York for the tall ships, and there was a state dinner for Queen Elizabeth and lots of heads of state." An eighteen-year-old was savoring history.

Susan, who now serves as chairman of the Betty Ford Foundation, found that the White House implanted lasting abstract influences: "It gave me incredible respect for the office of the President and the families who live there. And I learned that trust— credibility—is invaluable." She adds with a laugh, "And having lived there, you just have a whole different take when you watch the news or read the newspaper." She cheers the fact that each new First Family is different—"I don't think we want a cookie-cutter White House, but we have a hard time with change from one set of standards to another."

William Howard Taft had told friends that he wanted to be President only for what it would do for his children. Seizing that advantage, his older son, Robert, rapidly advanced up the Republican ladder, becoming Speaker of the House and then majority leader of the Senate—with the goal of becoming the first son since John

Quincy Adams to follow his father into the White House. Three times "Mr. Republican" competed for the GOP nomination for President, but in 1952, when it looked like a sure thing, along came General Eisenhower, who turned out to be a Republican—and the convention decided "We Like Ike." At Robert Taft's death, six months after Eisenhower's inauguration, he was given posthumous appreciation—a soaring bell tower on the Capitol grounds, the only such memorial to a member of Congress.

The youngest of the three Taft children, twelve-year-old Charlie, was the White House scamp, a former member of Quentin Roosevelt's mischievous White House Gang. Whatever would become of a boy who tied the guests' shoelaces together under the White House table? Quite a lot: Charles Phelps Taft served sixteen terms on the Cincinnati City Council and one term as mayor, had seven children, settled strikes, built low-cost housing, served as president of the international YMCA, and cofounded the World Council of Churches. The White House surely inspired such service, but he probably never revealed that he and Quentin once threw spitballs at Andrew Jackson's portrait.

While the Tafts' debutante daughter, Helen, was a tireless butterfly who greeted 1,208 guests at her coming-out tea in 1910 and another 300 at an East Room ball, the history around her was taking root within her. She became a distinguished professor of history and dean of Bryn Mawr College, married a Yale professor, had two daughters, and campaigned wholeheartedly for women's right to vote. Butterfly Helen metamorphosed into a trailblazer for the woman who can do it all. In what is rated an unsuccessful presidency, the White House produced lasting results in the next Taft generation.

NANCY REAGAN CALLS it "the very small sisterhood"; Luci Baines Johnson speaks of "the special relationship between all First Families, without regard to political party. There are so few of us in that fraternity who have lived in the President's House." Only thirty-nine First Ladies (including Martha Washington, who shared the life but not the house) and about a hundred presidential children have been White House families. They have loved it and hated it, have found it a palace and a prison. Most knew they were living amid history as it happened; others have taken longer to grasp the specialness of their lives.

There have been near incomprehensible changes in the role of First Ladies. Try to place Martha Washington in her paniers and mob cap in the same frame as Laura Bush lifting weights two or three times a week. The changes have generally come gradually, and the First Lady who dares to stretch the role one step further is still likely to get her knuckles cracked. Senator Clinton may test that undefined limit as she eyes the presidency in 2008.

But to see White House families laced together across the generations—and across political lines—has always brought a warm sense of continuity. First Lady Sarah Polk brings back Dolley Madison, a spry eighty, as the star at the Polks' parties; Maria Monroe, the first daughter of a President to marry in the mansion, writes a poem for the second, Lizzie Tyler, at her betrothal; Nellie Grant attends Alice Roosevelt's wedding, and Alice in turn attends the Johnson daughters' and Tricia Nixon's; General U. S. Grant III serves as military aide to Theodore Roosevelt; Harry Truman plays the piano at the Kennedys' dinner in his honor and drops in for lunch with the LBJs.

Grace Coolidge writes Bess Truman a letter of support, as one "accidental" First Lady to another; Jackie Kennedy writes Nancy Reagan a compassionate letter when President Reagan was shot, then telephones to add comfort. Patti Davis warns Chelsea Clinton, "A President's daughter should stay out of the spotlight." (Which Chelsea managed better than Patti.)

Three busy First Ladies—Laura Bush, Hillary Clinton, and Rosalynn Carter—appear jointly to benefit the Alzheimer's Association, and the one most affected, Nancy Reagan, who is ill, sends a touching letter. Four former Presidents and First Ladies attend the funeral of Ronald Reagan, number forty in the unbroken brotherhood who have shared the President's House. Three years later Nancy, wrapped in the mantle of her husband's magic, flies to Washington to plead the case for stem cell research—and changes the position of at least one powerful senator, Republican John McCain.

And then there is the ultimate continuity: Barbara and George Bush coming "home" to their son's White House.

To draw the tiny band of Presidents' children closer together, Lynda Johnson Robb gathered a number of them, of both parties, for a day of memories and reflections on their singular lives: "We wanted to share some of the things that only we could share," Lynda says. "We laughed together. . . . I told stories I have never told anybody, some that might be perceived as indiscreet." They could freely swap tales and personal thoughts about the Executive Mansion, knowing that none of this group, who had learned to treasure privacy, would repeat them. They might have grumbled about the demands and restrictions while they were there, but as adults they look back with appreciation. Luci Johnson, who arrived in the White House at sixteen and left five years later as a young mother, reflects: "It shaped my character—it was a prism through which I would view all other experiences of my life."

In his first days in the President's House that was his for only four miserable months, John Adams wrote to Abigail the words that are carved in the State Dining Room mantel today: "I pray heaven to bestow the best of blessings on this house, and on all that shall hereafter inhabit it. May none but honest and wise men ever rule under this roof."

Two centuries later, White House daughter Luci Baines Johnson speaks for those who have dwelled under that roof: "The impact is forever. . . . What extraordinary opportunity was mine!"

Notes

At the heart of the American republic is an eighteen-acre patch and a single house, and with the house comes a family—which is what this book is all about. Exploring the shared experiences of thirty-nine very individual First Ladies and forty-three families of varying composition has been a challenge, something like trying to put shoes on an octopus—these people were individuals who, though most are long gone, resisted my efforts to fit them into a group.

Much of the information about contemporary First Families comes from my direct observations, personal contacts, and interviews during many years with *Time* magazine. Books cited in the source listings that follow are but a fraction of those I consulted. I searched for books written in the nineteenth and early twentieth centuries, by men and women who worked in the White House or journalists who were present at the events they described. For recent times I turned again to books written by dependable White House employees, First Ladies' key aides, and journalists whose concern for precise and accurate reporting is highly regarded, a number of whom I know and respect.

One   Stepping into History

4   Jacqueline Kennedy had received invitations: Baldrige, *A Lady, First*, p. 166.

4   "I never wanted to be . . .": Cook, *Eleanor Roosevelt*, p. 472.

5   "I feel as if I am suddenly onstage . . .": Montgomery, *Mrs. LBJ*, p. 200.

6   "I felt like I'd been thrown into a river . . .": Trude Feldman, *McCall's*, Oct. 1974.

6   opposed Republican Party policy: *Today*, March 2004.

7   "I'd like it, because . . .": Mary Walker, Greenwich (Conn.) Library
         Oral History Project, 1991.

7   "Dad is looking . . .": Strober and Strober, *Ronald Reagan*, p. 40.

8   "It really is a little house": *Life*, Oct. 30, 1992.

8   "To go into those historic . . .": *Echoes of the White House* (Public
         Broadcasting Service documentary), 2001.

8   "that majestic office": Ibid.

8   "It's the best public . . .": White House Historical Association, *For
         Your Information*, p. 1, available at www.whitehousehistory.org.

9   "I sit here in this old . . .": McCullough, *Truman*, p. 398.

9   "danced a little jig": West, *Upstairs at the White House*, p. 110.

9   "I think of Lincoln . . .": Jensen, *The White House and Its Thirty-two
         Families*, p. 188.

10   complete inventory . . . : William Allman, interview.

10   Who was this fumbling his way: Aikman, *The Living White House*,
         p. 24.

10   Just two months earlier: Bassett, *Profiles and Portraits of American
         Presidents and Their Wives*, pp. 137–38.

11   In that vitriolic atmosphere: Poore, *Perley's Reminiscences of Sixty
         Years in the National Metropolis*, vol. 2, pp. 339–41.

13   Two weeks later, Abigail: Sadler, *America's First Ladies*, pp. 25–26.

14   dying of alcoholism: Jensen, *White House*, p. 9.

14   "Shiver, shiver . . . which is true": Ibid., p. 11.

15   "I know too much . . .": Sadler, *America's First Ladies*, p. 14.

16   "I mean, it's a huge platform": Laura Bush made these remarks at the
         National Press Club, Nov. 8, 2001.

16   "What a terrible responsibility . . .": Sadler, *America's First Ladies*,
         p. 129.

17   "The court . . . than as President": Bassett, *Profiles and Portraits*, p.
         256.

Two   "Out They Go, In We Come"

In recounting the instant turnover of White House families, nothing is as colorful and credible as the reminiscences of the new residents and the quite different ones of the long-serving staff. To these insiders' viewpoints, I have
added the perspective of White House correspondents, including myself, who
have observed many changeovers.

19 Luci Baines Johnson talked about the fire in her bedroom and other White House memories at the Lyndon Baines Johnson Presidential Library, Dec. 14, 2003.

21 "Hail to the Chief ": Poore, *Perley's Reminiscences of Sixty Years in the National Metropolis,* vol. 1, p. 511.

21 welcoming Texas into the Union: Jensen, *The White House and Its Thirty-two Families,* p. 67.

21 glittering social mecca: Poore, *Perley's Reminiscences,* vol. 1, p. 516.

22 "If you can't . . . counting man-hours": West, *Upstairs at the White House,* pp. 290–91.

23 "Now, that's the way . . .": Ibid., p. 75.

23 yellow bathing suit: Ibid., p. 30.

23 who stayed for twenty-one years: Parks, *My Thirty Years Backstairs at the White House,* p. 109.

24 "Of all the changes . . . swimming pool": West, *Upstairs,* p. 360.

25 Welsh terrier Charlie: Bryant, *Dog Days at the White House,* p. 121.

25 The change affected: Fields, *My 21 Years in the White House,* p. 28.

25 "Within a few days . . .": Ibid., p. 17.

26 the unbending John Quincy Adams: Poore, *Perley's Reminiscences,* vol. 1, p. 94.

27 "I remember looking out . . .": Strober and Strober, *Ronald Reagan,* p. 40.

30 "Well, Warren . . .": "First Ladies: Political Role and Public Image," Smithsonian Institution traveling exhibit, 2004.

31 Coolidge woke his father: Bassett, *Profiles and Portraits of American Presidents and Their Wives,* p. 302.

31 "maintain as far as possible . . .": Jensen, *White House,* p. 224.

32 "a home rich in tradition . . .": Bassett, *Profiles and Portraits,* p. 314.

33 "We are plain people . . .": Poore, *Perley's Reminiscences,* vol. 1, p. 205.

34 A weary messenger: Ibid., p. 269.

35 a special train that sped him: Bassett, *Profiles and Portraits,* p. 251.

35 "President McKinley died . . . without fear for Theodore": Morris, *Edith Kermit Roosevelt,* pp. 212–14.

Three   Growing Up in Headlines

From the earliest days of the republic, Americans have been interested in—not to mention nosy and judgmental about—the children who live in the White House. The children's chronicles began with Jefferson's seven grandchildren, the first of dozens who have brightened life in the President's House over two centuries. (John and Abigail Adams's little granddaughter was there for less than three months.)

Everything written about and by a First Lady naturally includes her children, and biographies of a President, at the very least, usually mention his family. The daughters have shared their singular role in articles, interviews, and a few books; presidential sons have mostly kept their inside-the-mansion stories to themselves. The sheer number of sources I consulted makes it impossible to cite all.

Four authors provided particularly rich and reliable material: the ever-dependable Ben: Perley Poore (*Perley's Reminiscences of Sixty Years in the National Metropolis,* 1886); Amy La Follette Jensen (*The White House and Its Thirty-two Families,* 1958, updated in 1962); Lonnelle Aikman (*The Living White House,* part of a series produced by the White House Historical Association with the National Geographic Society, 1966); and Christine Sadler (*Children in the White House,* 1967). In using the recollections of White House staff members, I have included their names in the text.

As a White House correspondent, I wrote about the First Children whenever they stuck their heads above the parapet and made news, interviewed the brides, covered the weddings, and watched them mature into admirable citizens. My extensive notes, including my interviews with First Ladies and White House daughters, were revisited for this chapter.

These sources provided specific points:

40   Icelandic ponies, though tiny: Morris, *Edith Kermit Roosevelt,* p. 124.

40   They "borrowed" the biggest metal trays: Longworth, *Crowded Hours,* p. 45.

42   Discipline was what Miss Arnold: Morris, *Edith Kermit Roosevelt,* p. 315.

44   "a slow-developing child . . .": Matthew Pinsker, "The Lincolns' Summer Place," *Preservation,* Sept.–Oct. 2003, p. 5.

45   Pert thirteen-year-old Nellie: Poore, *Perley's Reminiscences of Sixty Years in the National Metropolis,* vol. 2, pp. 251–52.

45   all-night dances, called "Germans": Sadler, *Children in the White House,* p. 179.

46   "You've got to protect Chelsea . . .": Hillary Clinton, " 'You've Got to Protect Chelsea at All Costs,' " *Time,* June 16, 2003, p. 36.

48   "Living in the White House . . .": *Life,* Oct. 30, 1992, p. 41.

58   Her life took an incredible turn: Sadler, *Children in the White House,* p. 247.

62   "Bear in mind . . .": Smith and Durbin, *White House Brides*, p. 57.

62   For Steve Ford and Barbara Walters, see Weidenfeld, *First Lady's Lady*, p. 77.

64   The worst of this lot: Sadler, *Children in the White House*, pp. 64–67.

Four   Love in the Fishbowl

Love and its evil twin, gossip, escalate beyond chitchat columns into news coverage when either or both occur in the White House. The volume of newspaper and magazine sources prohibits my listing them all.

68   For the reaction to Julia Gardiner, see Jensen, *The White House and Its Thirty-two Families*, p. 66.

68   Paranoid that the venue: Bassett, *Profiles and Portraits of American Presidents and Their Wives*, p. 105.

68   What sort of gift: Jefferson Rarities catalog, p. 19.

69   Frances disembarked: Barry, *Forty Years in Washington*, p. 166.

70   ". . . I am to be married . . .": Poore, *Perley's Reminscences of Sixty Years in the National Metropolis*, vol. 2, pp. 516–20.

75   "Well, for Mr. Jackson's sake . . .": Ibid., vol. 1, p. 92.

76   divorce had not been finalized: Bassett, *Profiles and Portraits*, p. 71.

76   a political millstone: Ibid., p. 81.

76   "May God Almighty forgive . . .": Sadler, *America's First Ladies*, p. 57.

77   Apparently quite content: Hope Ridings Miller, *Scandals in the White House*, p. 150.

77   The gossips were unaware: Poore, *Perley's Reminiscences*, vol. 2, p. 437.

77   stained-glass window: St. John's Church brochure.

78   "The President looked . . . station house with you": Butt, *Taft and Roosevelt*, p. 98.

79   "My wife . . . be careful how you tell her": Jensen, *Profiles and Portraits*, p. 178.

82   his indomitable mother, Sara: Goodwin, *No Ordinary Time*.

84   exception to his social frostbite: Bloom, *There's No Place Like Washington*, p. 20.

84   "I think you will get along . . .": Furman, *White House Profile*, p. 321.

84   Grace went for a walk: Starling, *Starling of the White House*, p. 252.

84   "the sunshine and joy . . .": Sadler, *America's First Ladies*, p. 199.

85   fortieth wedding anniversary: Ibid., p. 244.

86   "They are woefully . . . but once been consulted": Allgor, *Parlor Pol-*
          itics, In Which the Ladies of Washington Help Build a City and a Gov-
          ernment, p. 157.

87   "I pray you come on immediately . . .": *Journal of the White House*
          Historical Association, Nov. 2002.

89   "She was forced . . . the White House blues": Truman, *Bess W. Tru-*
          man, pp. 253–70.

Five   Whose Life Is It Anyway?

Privacy—when is the First Family a legitimate subject, when not?—has been
an issue in the White House since its earliest days, and it only becomes more so
as the media grows ever larger and its tools more sophisticated. Still, the White
House needs the media for its own ends.

93   "One hates to feel . . .": Morris, *Edith Kermit Roosevelt,* p. 205.

95   "not the least fence . . .": Aikman, *The Living White House,* p. 26.

95   "I was somewhat annoyed . . .": Grant, *The Personal Memoirs of Julia*
          Dent Grant (Mrs. Ulysses Grant), p. 174.

96   Long before his wedding: Barry, *Forty Years in Washington,* p. 168.

96   the Clevelands carved out a unique life: Bassett, *Profiles and Portraits*
          of American Presidents and Their Wives, p. 216.

97   the President climbed into his carriage: Barry, *Forty Years,* p. 169.

97   President Hoover looked up from his dinner: Starling, *Starling of the*
          White House, p. 285.

98   And then there was the cheeky couple: Cook, *Eleanor Roosevelt,* p.
          472.

100   "The hardest part is that I wish . . .": Strober and Strober, *Ronald*
          Reagan, p. 45.

101   "I'm just going to be . . .": Cook, *Eleanor Roosevelt,* p. 476.

103   Even Jackie Kennedy: Baldrige, *A Lady, First,* p. 179.

105   Luci Baines Johnson and Lynda Johnson Robb spoke about their
          disguises at Marquette University and during a White House tour,
          respectively, at the Lyndon Baines Johnson Presidential Library,
          Dec. 14, 2003, and in author interviews.

114   Imagine today's photographers: Clapper, *Washington Tapestry,* p. 113.

116   Women reporters: Poore, *Perley's Reminscences of Sixty Years in the*
          National Metropolis, vol. 2, p. 106.

Six All in the Family

Over the years the families have been as different as their President-fathers. This chapter searches out their lives within the White House and the impact the White House has had on those lives when they step outside its fences. While much is known about contemporary children—to their dismay—the nineteenth-century accounts give us a glimpse into a quieter world of family, yet even then troubled by an intruding public.

119   "There is something in the great unsocial house . . .": Boller, *Presidential Wives*, p. 53.

122   "He could see into our hearts . . .": Sadler, *Children in the White House*, pp. 56–57.

122   old soldier played Cupid: Smith and Durbin, *White House Brides*, p. 45.

122   His gift for baby Mary Rachel: Sadler, *Children in the White House*, p. 94.

123   a mansion that had been bought for the general: Bassett, *Profiles and Portraits of American Presidents and Their Wives*, p. 175.

124   Driving his spirited team: Collins, *Presidents on Wheels*, p. 89.

125   323 houseguests: Aikman, *The Living White House*, p. 54.

125   "the soft side of the billiard table": Sadler, *Children in the White House*, p. 185.

126   her family made themselves at home: Butt, *Taft and Roosevelt*, p. 15.

126   waltz and two-step: Smith, *Entertaining in the White House*, p. 164.

126   Ever-practical Nellie: Bassett, *Profiles and Portraits*, p. 269.

127   "The White House is not . . .": Sidey, *A Very Personal Presidency*, p. 47.

128   "one apple, sliced thin": Barry, *Forty Years in Washington*, p. 289.

129   Soon a specially constructed bathtub: Jensen, *The White House and Its Thirty-two Families*, p. 197.

129   Laddie Boy: Starling, *Starling of the White House*, p. 184.

130   "hard-fought games with mallet and ball": Aikman, *The Living White House*, p. 136.

130   Coolidge took up riding: Starling, *Starling of the White House*, p. 124.

131   Supreme Court Justice Harlan Stone: William E. Leuchtenberg quoted in Jeffrey Rosen, "The Justice Who Came to Dinner," *New York Times*, Feb. 1, 2004.

132   Eleanor's dog, Major: Furman, *Washington By-Line*, p. 165.

132   He astonished guests: Smith, *Entertaining in the White House*, p. 180.

133   "What about the appropriation?": Parks, *My Thirty Years Backstairs at the White House,* p. 179.

135   "I wonder if you will ever . . .": Freidel, *The Presidents of the United States,* p. 91.

136   "It's kind of fun . . .": David Eisenhower, C-Span interview, Oct. 2, 2003.

138   over a plate of beans: Matthew Pinsker, "The Lincolns' Summer Place," *Preservation,* Sept.–Oct. 2003, pp. 55–57.

Seven   The Mirror of the Times

Facts about the construction of the White House appear in many sources. Bits and pieces about some facet of this broad view of the White House as a reflection of the times are to be found in more works than can be listed. Certain references are widely used; those below were particularly useful.

140   "imposing . . . In size . . .": Aikman, *The Living White House,* p. 25.

140   Along with fellow Virginian: Ibid., p. 132.

140   "He even nudged . . .": Forrest McDonald, *New York Times Book Review,* Nov. 7, 2004, p. 18.

140   "We want nothing here . . .": Brookhiser, *Gentleman Revolutionary,* pp. 162–63.

141   "What will the common people . . .": John C. Miller, *The Federalist Era,* p. 8.

142   "First Lady" initially appeared: Kane, *Facts About the Presidents,* p. 399.

143   The cornerstone was finally laid: Jensen, *The White House and Its Thirty-two Families,* p. 6.

146   Dolley's travails were the stuff: Bassett, *Profiles and Portraits of American Presidents and Their Wives,* p. 44.

147   In 1842 Samuel Morse: Poore, *Perley's Reminiscences of Sixty Years in the National Metropolis,* vol. 1, p. 310.

147   For the advent of gaslights, bathtubs, cookstoves, and telephones, see Aikman, *The Living White House,* pp. 120–22.

148   In 1891 electricity: Bassett, *Profiles and Portraits,* p. 226.

149   "neither ice nor fire . . .": Aikman, *The Living White House,* pp, 107–8.

152   A special commission offered three alternatives: Seale, *The White House Papers,* vol. 2, p. 1029.

153   For details on White House life during World War II, see West, *Upstairs at the White House,* pp. 34–37.

154   The First World War had been different: Jensen, *White House,* p. 208.

154   For Margaret Wilson entertaining the troops, see Sadler, *Children in the White House,* p. 234.

155   "It was as in Paris . . . honest-looking coats": Poore, *Perley's Reminiscences,* vol. 2, pp. 111–12.

Eight   White House à la Mode

Until Jackie Kennedy stabilized the White House decor, a new First Lady's re-furbishing could set the style for the nation's homemakers—until a successor came in with a new idea of decorating. And across the years First Ladies and their well-connected guests have been a living catalog of the latest in fashion—especially in the nineteenth century, when dressing up was very serious busi-ness. From Martha Washington to Laura Bush, a First Lady's every outfit is critiqued, most trenchantly by the observant Perley Poore.

158   "I was disappointed . . .": Baldrige, *A Lady, First,* p. 177.

159   "Everything in the White House . . .": Sidey, *A Very Personal Presidency,* p. 33.

160   the plight of John Tyler: Jensen, *The White House and Its Thirty-two Families,* p. 63.

160   "a badly kept barracks . . .": Smith, *Entertaining in the White House,* p. 127.

161   "He sold a pair . . .": Merle Miller, *Plain Speaking,* p. 140.

161   "it was high time . . .": Nesbitt, *White House Diary,* p. 176.

162   At the bargain price: Jensen, *White House,* p. 39.

162   "No matter what they build . . .": Poore, *Perley's Reminiscences of Sixty Years in the National Metropolis,* vol. 2, p. 350.

164   Lou Hoover, who shuffled: Parks, *My Thirty Years Backstairs at the White House,* p. 206.

164   "Now I know why . . .": West, *Upstairs at the White House,* p. 62.

166   "Hell itself couldn't warm . . .": Aikman, *The Living White House,* p. 171.

166   "Surely, the greatest brains . . .": West, *Upstairs,* p. 200.

168   "You will feel that you are . . .": Shelton, *The White House Today and Yesterday,* p. 124.

168   a bird-of-paradise plume: Jensen, *White House,* p. 20.

169   "in danger of being confounded . . .": Smith, *Entertaining,* p. 47.

171   The Johnson ladies, ever aware: Poore, *Perley's Reminiscences,* vol. 2,
           p. 330.
171   "Milliners sent . . .": Smith, *Entertaining,* p. 47.
171   Mamie Eisenhower's closets: Parks, *My Thirty Years,* p. 331.
174   Chester A. Arthur is the clear winner: Jensen, *White House,* p. 130.

Nine   Twenty-nine Courses,
Four Thousand Hands to Shake

Official entertaining has been a major element of White House life, and from
the first receptions until today the public has shown an appetite for details.
While this chapter draws from many sources—not least my own years of cov-
ering many as a reporter and attending a number as a guest—the most com-
plete accounts are found in Marie Smith's *Entertaining in the White House.*
Perley Poore's colorful reports, as always, bring the nineteenth-century White
House alive; Lonnelle Aikman's *The Living White House,* published by the
White House Historical Association, digs into early social records; and in the
section on Blair House, President Reagan's chief of protocol, Selwa "Lucky"
Roosevelt, is the most knowledgeable and amusing source.

The following writers provided specific points about the semipublic side of
the President's House.

177   "I will not permit . . .": Aikman, *The Living White House,* p. 53.
178   The concert was a close call: Baldrige, *A Lady, First,* p. 197.
179   "When the Prince . . .": Cassini, *Never a Dull Moment,* p. 169.
181   "More is to be performed . . .": Smith, *Entertaining in the White
           House,* p. 30.
181   standing on a raised platform: Aikman, *The Living White House,* p. 45.
185   For the Lincolns' inaugural reception, see Smith, *Entertaining,* p. 97.
187   The anecdote about the Kennedy-era head of state comes from ibid.,
           p. 260.
193   "I am glad this is . . .": Holloway, *Ladies of the White House,* p. 627.
193   "4,100 pieces . . .": Head Usher Gary Walters, interview.
194   In Lincoln's day: Aikman, *The Living White House,* p. 92.
194   Silver spoons were small potatoes: Nesbitt, *White House Diary,* p. 89.
195   "Remove the tray quietly . . .": Fields, *My 21 Years in the White House,*
           p. 40.
195   the tail of Old Whitey: Jensen, *The White House and Its Thirty-two
           Families,* p. 37.

196   At breakfast, scotch: Fields, *My 21 Years*, p. 51.

196   For Foreign Minister Molotov's luggage, see West, *Upstairs at the White House*, p. 44.

198   Many requirements were persnickety: Nesbitt, *White House Diary*, pp. 222–27.

198   For the demanding royal servants, see Fields, *My 21 Years*, p. 44.

200   King David Kalakaua: Smith, *Entertaining*, p. 111.

201   two wives of the King of Nepal?: *Life*, Oct. 30, 1992, p. 25.

204   ". . . at home in Buffalo": Poore, *Perley's Reminiscences of Sixty Years in the National Metropolis*, vol. 1, p. 386.

205   her entourage of forty: West, *Upstairs*, p. 44.

Ten   Intramural Skirmishes

Since this chapter is made up of eight self-contained episodes, the principal sources are, in general, contained in their one section. The following books were singled out as the most personal accounts of the turbulent events; four of the authors wrote as personal witnesses.

207   The fracas erupted: Poore, *Perley's Reminiscences of Sixty Years in the National Metropolis*, pp. 120–24. Poore was on the scene for President Andrew Jackson's showdown with his cabinet over Peggy Eaton.

208   As the White House social issue: Allgor, *Parlor Politics, In Which the Ladies of Washington Help Build a City and a Government*, pp. 204–9.

209   After being "decontaminated": Hope Ridings Miller, *Scandals in the White House*, pp. 118–20, 128.

209   Anthony Merry: Jensen, *The White House and Its Thirty-two Families*, pp. 17–18.

209   a cultivated Virginian and Francophile: Brodie, *Thomas Jefferson*, p. 556.

210   The Merrys, seething, stalked out: Allgor, *Parlor Politics*, p. 37.

211   The young Mrs. Hay: Jensen, *White House*, pp. 17–18.

212   "The drawing room of the President . . .": Sadler, *Children in the White House*, p 72.

213   "earthquake, upheaval and cyclone": Andrews, *My Studio Window*, p. 368.

214   "Hostesses were afraid . . .": Bloom, *There's No Place Like Washington*, pp. 43–46.

215   "Protocol is as much the law . . .": Ibid., p. 43.

215   a Solomonesque solution: Smith, *Entertaining in the White House,* p. 191.

216   "looked always as if they'd caught . . .": McCullough, *Truman,* pp. 593–94.

217   For the proposed 1889 renovation, see Bassett, *Profiles and Portraits of American Presidents and Their Wives,* p. 226.

220   "Mrs. Roosevelt and Franklin Jr. . . .": Merle Miller, *Plain Speaking,* p. 227.

Eleven   Slings and Arrows

Though the mansion is grand and the title distinctive, neither is a shield against the criticism that inevitably comes with both. Sometimes the censure is unfair, sometimes not—depending upon the viewpoint—and most of the dust-ups make news. The sources listed here provide little-known incidents or colorful details about those that made headlines.

223   "It is not good form . . .": Bassett, *Profiles and Portraits of American Presidents and Their Wives,* p. 130.

226   "I saw Cousin Eleanor . . .": Archie Roosevelt, *For Lust of Knowing,* pp. 40–42.

226   When Lash was drafted: West, *Upstairs at the White House,* p. 37.

231   "he did throw a pillow at me": Wiedenfeld, *First Lady's Lady,* p. 172.

232   a headstrong First Lady: Baldrige, *A Lady, First,* pp. 199–200.

234   "I must dress in costly . . .": Smithsonian Institution traveling exhibition.

235   "The Queen Must Dance": Poore, *Perley's Reminiscences of Sixty Years in the National Metropolis,* vol. 2, p. 120.

235   When it comes to vituperation: Bassett, *Profiles and Portraits,* p. 301.

236   a billiards table: Poore, *Perley's Reminiscences,* vol. 1, p. 31.

237   "extremely attractive . . . finest singers": McCullough, *Truman,* p. 828.

237   the President exploded: Merle Miller, *Plain Speaking,* pp. 87–88.

237   mail ran two to one: McCullough, *Truman,* p. 567.

238   For turkeys in the White House, see Barry, *Forty Years in Washington,* p. 269.

239   "Any man familiar . . .": Hope Ridings Miller, *Scandals in the White House,* pp. 3–4.

239   "I could wish the women . . .": Ibid., p. 164.

241   "He dwelt on the gorgeous . . . proudest Asiatic mansion?": Poore, *Perley's Reminiscences,* vol. 1, p. 228.

Twelve   Dark Days

Some Presidents, notably William Henry Harrison, Zachary Taylor, and James Garfield, are known more for their deaths than for their accomplishments; the three of them combined served less than two years in the White House, hardly generating enough action for biographies, much less memoirs.

Volumes have been written about the killing of President Lincoln, but one that I found to have particular immediacy is *Life and Times of Abraham Lincoln* (especially pp. 611–76), written by a friend of the President, L. P. Brockett, and published in 1865, only months after the assassination. He drew from official depositions that reconstructed every move, beginning with Major Henry R. Rathbone, the Lincolns' guest in the Ford theater box. Brockett's descriptions of the White House in mourning and his observations aboard the funeral train on its long journey back to Illinois are compelling.

At Taylor's death, Perley Poore was once again there to cover the story (*Perley's Reminiscences of Sixty Years in the National Metropolis,* vol. 1, p. 418). A report of the exhuming of the twelfth President—to settle lingering suspicions that he might have been poisoned—appeared 142 years later in *Life* (Oct. 30, 1992).

248   caught one of the many rabbits: Sadler, *Children in the White House,* p. 254.

254   Henry Wallace: Robbins, *Bess and Harry,* p. 79.

254   "I was very apprehensive . . .": Ibid., p. 78.

257   "I should think . . .": Rossiter, *The American Presidency,* p. 203.

257   Alexander Graham Bell rushed: White House Historical Association website, www.whitehousehistory.org.

258   "Oh, why am I made . . .": Bassett, *Profiles and Portraits of American Presidents and Their Wives,* p. 198.

259   "You, especially . . .": Jacqueline Kennedy's farewell note to her son is quoted in Christopher Andersen, *The Day John Died,* p. 207.

Thirteen   Bittersweet Farewells

A First Family's manner of leaving the White House tells much about their happiness as residents. In one of her many compelling letters, Abigail Adams declared, "I am sick, sick and sick of publick life" (Richard S. Ryerson, ed., *Adams*

Family Correspondence); at the other extreme, Theodore Roosevelt told his aide and friend Archie Butt, "I have enjoyed every moment" (Butt, *Taft and Roosevelt*).

For personal reminiscences about leaving the President's House, the memoirs of a number of First Ladies provide their unique view, particularly Helen Taft's *Recollections of a Full Life*, Lady Bird Johnson's *A White House Diary*, and Betty Ford's *The Times of My Life*.

Among White House employees, both Lillian Parks, in *My Thirty Years in the White House*, and J. B West, in *Backstairs at the White House*, reflect the sharp differences in reactions to departing families—sometimes tearful good-byes, other times smiles of relief.

262   His last act before leaving: Merle Miller, *Plain Speaking*, p. 10.

263   "plain Mr. Truman now . . .": McCullough, *Truman*, p. 992.

264   ". . . I feel a sense of awe . . .": *Echoes of the White House* (Public Broadcasting Service documentary), 2001.

265   "My election as President . . .": Bassett, *Profiles and Portraits of American Presidents and Their Wives*, p. 61.

267   "I was bitter . . .": *Today*, Feb. 1, 2004.

267   "The nearer I get . . .": Jensen, *The White House and Its Thirty-two Families*, p. 198.

268   "I am heartily tired . . .": Kane, *Facts About the Presidents*, p. 165.

269   "If you are as happy . . .": Ibid., p. 181.

270   For Frances Cleveland as a dinner guest of the Tafts, see Smith, *Entertaining in the White House*, p. 164.

273   a terrible jolt: Kane, *Facts About the Presidents*, p. 222.

275   For the Eisenhowers' farewell evening, see Smith, *Entertaining*, p. 244.

276   "Never did a prisoner . . .": Jensen, *White House*, p. 20.

Fourteen   Footprints in History

The closing chapter wraps up my notes, my opinions, and, best of all, my memories of the First Families I have observed and known—or feel that I know after delving into their lives in the White House. They reflect the nation as it changes; the nation has been fortunate to have such shining mirrors.

Bibliography

Books

Aikman, Lonnelle. *The Living White House*. White House Historical Association, 1966.

Alderman, Ellen, and Caroline Kennedy. *The Right to Privacy*. Alfred A. Knopf, 1995.

Allgor, Catherine. *Parlor Politics, In Which the Ladies of Washington Help Build a City and a Government*. University Press of Virginia, 2000.

Ames, Mary Clemmer. *Ten Years in Washington: Life and Scenes in the National Capital*. A.D. Worthington, 1878.

Andersen, Christopher. *The Day John Died*. William Morrow, 2000.

———. *Sweet Caroline*. William Morrow, 2003.

Andrews, Marietta Minnigerode. *My Studio Window: Sketches of the Pageant of Washington Life*. E.P. Dutton, 1928.

Baldrige, Letitia. *A Lady, First: My Life in the Kennedy White House and the American Embassies of Paris and Rome*. Viking, 2001.

Bannett, Carole. *Partners to the President: President Johnson's Wife and Daughters Speak*. Citadel, 1966.

Barker, Charles E. *With President Taft in the White House*. A. Kroch and Son, 1947.

Barry, David S. *Forty Years in Washington*. Little, Brown, 1924.

Bassett, Margaret. *Profiles and Portraits of American Presidents and Their Wives*. Bond Wheelwright, 1964; reprint, 1969.

Benson, Harry. *The President and Mrs. Reagan*. Harry N. Abrams, 2003.

Berquist, Laura, and Stanley Tretick. *A Very Special President*. McGraw-Hill, 1965.

Blackman, Ann. *Wild Rose: Civil War Spy, a True Story.* Random House, 2005.

Bloom, Vera. *There's No Place Like Washington.* G.P. Putnam's Sons, 1944.

Boller, Paul F., Jr. *Presidential Wives: An Anecdotal History.* Oxford University Press, 1988.

Bowen, Ezra. *This Fabulous Century: Sixty Years of American Life.* Time-Life, 1969.

Brands, H. W. *T.R.: The Last Romantic.* Basic Books, 1997.

Brockett, L. P. *Life and Times of Abraham Lincoln.* Bradley, 1865.

Brodie, Fawn M. *Thomas Jefferson: An Intimate History.* W.W. Norton, 1974.

Brookhiser, Richard. *Gentleman Revolutionary: Gouverneur: The Rake Who Wrote the Constitution.* Free Press, 2003.

Bryant, Traphes, with Frances Spatz Leighton. *Dog Days at the White House.* Macmillan, 1975.

Buckley, Christopher. *Washington Schlepped Here: Walking in the Nation's Capital.* Crown, 2003.

Burt, Nathaniel. *First Families: The Making of an American Aristocracy.* Little, Brown, 1970.

Bush, Barbara. *Barbara Bush: A Memoir.* Lisa Drew/Charles Scribner's Sons, 1994.

Bush, George H. W. *All the Best, George Bush: My Life in Letters and Other Writings.* Lisa Drew/Scribner, 1999.

Bush, George W. *A Charge to Keep.* William Morrow, 1999.

Butt, Archibald W. *Taft and Roosevelt: The Intimate Letters of Archie Butt, Military Aide.* Doubleday, Doran, 1930.

Caroli, Betty Boyd. *First Ladies.* Oxford University Press, 1987; reprint, 1995.

Carpenter, Liz. *Ruffles and Flourishes.* Doubleday, 1970.

Carter, Rosalynn. *First Lady from Plains.* Houghton Mifflin, 1984.

Cassini, Countess Marguerite. *Never a Dull Moment.* Harper & Brothers, 1936.

Clapper, Olive Ewing. *Washington Tapestry.* McGraw-Hill, 1946.

Clinton, Hillary Rodham. *Living History.* Simon & Schuster, 2003.

Colacello, Bob. *Ronnie and Nancy: Their Path to the White House, 1911 to 1980.* Warner Books, 2004.

Collins, Herbert Ridgeway. *Presidents on Wheels.* Acropolis, 1971.

Colman, Edna M. *Seventy-five Years of White House Gossip, from Washington to Lincoln.* Doubleday, Page, 1926.

Cook, Blanche Wiesen. *Eleanor Roosevelt.* Viking, 1992.

Crook, Colonel W. H. *Through Five Administrations.* Harper & Brothers, 1910.

————. *Memories of the White House: The Family Life of Our Presidents from Lincoln to Roosevelt*. Little, Brown, 1911.

Davis, Patti. *The Way I See It*. G.P. Putnam's Sons, 1992.

Deaver, Michael. *Nancy: A Portrait of My Years with Nancy Reagan*. William Morrow, 2004.

Durant, John, and Alice Durant. *Pictorial History of American Presidents*. A.S. Barnes, 1955.

————. *The Presidents of the United States*. Vols. 1 and 2. A.A. Gache and Son, 1976.

Eisenhower, Julie Nixon. *Special People*. Simon & Schuster, 1977.

————. *Pat Nixon: The Untold Story*. Simon & Schuster, 1986.

Eisenhower, Susan. *Mrs. Ike*. Farrar, Straus & Giroux, 1996.

Ellis, Joseph. *American Sphinx: The Character of Thomas Jefferson*. Random House, 1996.

Ferling, John. *Adams vs. Jefferson: The Tumultuous Election of 1800*. Oxford University Press, 2004.

Fields, Alonzo. *My 21 Years in the White House*. Fawcett, 1960.

Ford, Betty, with Chris Chase. *The Times of My Life*. Harper & Row/Reader's Digest Press, 1979.

————. *Betty: A Glad Awakening*. Doubleday, 1987.

Freidel, Frank. *The Presidents of the United States*. White House Historical Association/National Geographic Society, 1964.

Furgurson, Ernest B. *Freedom Rising: Washington in the Civil War*. Alfred A. Knopf, 2004.

Furman, Bess. *Washington By-line*. Alfred A. Knopf, 1949.

————. *White House Profile*. Bobbs Merrill, 1951.

Gerber, Robin. *Leadership the Eleanor Roosevelt Way*. Prentice-Hall, 2002.

Gerhart, Ann. *The Perfect Wife: The Life and Choices of Laura Bush*. Simon & Schuster, 2004.

Gibson, Barbara, and Ted Schwartz. *Rose Kennedy and Her Family*. Birch Lane/Carol, 1995.

Goodwin, Doris Kearns. *No Ordinary Time*. Simon & Schuster, 1994.

Grant, Julia Dent. *The Personal Memoirs of Julia Dent Grant (Mrs. Ulysses S. Grant)*. Edited, with notes by John Y. Simon. Southern Illinois University Press, 1975.

Green, Constance. *Washington: Capital City, 1879–1950*. Vol. 2. Princeton University Press, 1962.

Hess, Stephen. *America's Political Dynasties*. Doubleday, 1966.

Hickey, Donald R. *The War of 1812: A Forgotten Conflict*. University of Illinois Press, 1990.

Holloway, Laura Carter. *The Ladies of the White House.* U.S. Publishing/H.H. Bancroft, 1870.

Hoover, Irwin H. *Forty-two Years in the White House.* Houghton Mifflin, 1934.

Jaffray, Elizabeth. *Secrets of the White House.* Cosmopolitan, 1926.

Jensen, Amy La Follette. *The White House and Its Thirty-two Families.* McGraw Hill, 1958; rev. eds., 1962, 1965.

Johnson, Haynes, with photographs by Frank Johnston. *The Working White House.* Praeger, 1975.

Johnson, Lady Bird. *A White House Diary.* Holt, Rinehart and Winston, 1970.

Johnson, Walter. *1600 Pennsylvania Avenue: Presidents and the People, 1929–1959.* Little, Brown, 1960.

Kane, Joseph. *Facts About the Presidents.* Pocket, 1993.

Keckley, Elizabeth Hobbs. *Behind the Scenes.* G.W. Carleton, 1868.

Kelley, Virginia Clinton, with James Morgan. *Leading with My Heart: My Life.* Simon & Schuster, 1994.

Kennedy, Caroline. *Profiles in Courage for Our Time.* Hyperion, 2003.

———. *A Patriot's Handbook.* Hyperion, 2003.

Kerney, James. *The Political Education of Woodrow Wilson.* Century, 1926.

Klapthor, Margaret Brown. *Dresses of the First Ladies of the White House.* Smithsonian Institution, 1952.

———. *Official White House China.* Smithsonian Instition, 1975.

Korda, Michael. *Ulysses S. Grant: The Unlikely Hero.* HarperCollins, 2004.

La Carruba, Michael L. *Presidents and Wives: Portraits, Facts.* Historic Publications, 1959.

Latimer, Louise Payson. *Your Washington and Mine.* Charles Scribner's Sons, 1924.

Leech, Margaret. *Reveille in Washington.* 1941; reprint, Simon Publications, 2003.

Leeming, Joseph. *The White House in Picture and Story.* George W. Stewart, 1953.

Leish, Kenneth W., ed. *The American Heritage Pictorial History of the Presidents of the United States.* Vols. 1, 2, and 3. American Heritage, 1968.

Levin, Phyllis Lee. *Edith and Woodrow: The Wilson White House.* Lisa Drew/Scribner, 2001.

Longworth, Alice Roosevelt. *Crowded Hours.* Charles Scribner's Sons, 1933.

Lorant, Stefan. *Lincoln: A Picture Story of His Life.* W.W. Norton, 1952.

Lowe, Jacques. *Kennedy: A Time Remembered.* Quartet/Visual Arts, 1983.

Lowe, Jacques; text by Hugh Sidey. *Remembering Jack: Intimate and Unseen Photographs of the Kennedys.* Bulfinch, 2003.

McAdoo, Eleanor Wilson. *The Woodrow Wilsons*. Macmillan, 1937.

McCaffree, Mary Jane, and Pauline Innis. *Protocol: The Complete Handbook of Diplomatic, Official and Social Usage*. Prentice-Hall, 1977.

McCullough, David. *Truman*. Touchstone/Simon & Schuster, 1992.

———. *John Adams*. Simon & Schuster, 2001.

Maraniss, David. *First in His Class: A Biography of Bill Clinton*. Touchstone/Simon & Schuster, 1996.

Means, Marianne. *The Women in the White House: The Lives, Times and Influence of Twelve Notable First Ladies*. Random House, 1963.

Miller, Hope Ridings. *Scandals in the Highest Office*. Random House, 1973.

Miller, John C. *The Federalist Era, 1789–1801*. Waveland Press, 1999.

Miller, Merle. *Plain Speaking: An Oral Biography of Harry S. Truman*. Berkley, 1974.

Miller, Nathan. *F.D.R.: An Intimate History*. Doubleday, 1983.

Minutaglio, Bill. *First Son: George W. Bush and the Bush Family Dynasty*. Times Books, 1999.

Montgomery, Ruth. *Mrs. LBJ*. Holt, Rinehart and Winston, 1964.

———. *Hail to the Chiefs: My Life and Times with Six Presidents*. Coward-McCann, 1970.

Morris, Roger. *Partners in Power*. John Macrae/Henry Holt, 1996.

Morris, Roy Jr. *Fraud of the Century: Rutherford B. Hayes, Samuel Tilden, and the Stolen Election of 1876*. Simon & Schuster, 2003.

Morris, Sylvia Jukes. *Edith Kermit Roosevelt: Portrait of a First Lady*. Coward, McCann & Geohegan, 1980.

Nagel, Paul C. *The Adams Women: Abigail and Louisa, Their Sisters and Daughters*. Harvard University Press, 1987.

Nesbitt, Henrietta. *White House Diary*. Doubleday, 1948.

Parks, Lillian Rogers. *My Thirty Years Backstairs at the White House*. Fleet, 1961.

Pearce, Mrs. John N. *The White House: An Historic Guide*. White House Historical Association/National Georgraphic Society, 1962; rev. ed., expanded by William V. Elder III and James R. Ketchum, 1964.

Perkins, Frances. *The Roosevelt I Knew*. Viking, 1946.

Perry, Mark. *Grant and Twain: The Story of a Friendship That Changed America*. Random House, 2004.

Pinsker, Matthew. *Lincoln's Sanctuary: Abraham Lincoln and the Soldiers' Home*. Oxford University Press, 2003.

Poore, Ben: Perley. *Perley's Reminiscences of Sixty Years in the National Metropolis*. Vols. 1 and 2. Hubbard Brothers, 1886.

Post, Robert C., ed. *Every Four Years: The American Presidency.* Smithsonian Exposition Books, 1980.

Radcliffe, Donnie. *Simply Barbara Bush.* Warner Books, 1989.

———. *Hillary Rodham Clinton: A First Lady for Our Time.* Warner Books, 1993.

Randolph, Mary. *Presidents and First Ladies.* Appleton-Century, 1936.

Randolph, Sarah H. *The Domestic Life of Thomas Jefferson.* Harper & Brothers, 1871.

Robbins, Jhan. *Bess and Harry: An American Love Story.* G.P. Putnam's Sons, 1980.

Reagan, Maureen. *First Father, First Daughter: A Memoir.* Little, Brown, 1989.

Reagan, Michael, with Joe Hyams. *On the Outside Looking In.* Kensington, 1988.

Reagan, Nancy, with William Novak. *My Turn.* Random House, 1989.

Roosevelt, Archie. *For Lust of Knowing: Memoirs of an Intelligence Officer.* Little Brown, 1988.

Roosevelt, Eleanor. *This Is My Story.* Garden City, 1937.

———. *This I Remember:* Harper & Brothers, 1961.

Roosevelt, Kermit. *Quentin Roosevelt.* Charles Scribner's Sons, 1921.

Roosevelt, Selwa. "Lucky." *Keeper of the Gate.* Simon & Schuster, 1990.

Ryerson, Richard Alan. *Adams Family Correspondence.* Ser. 2. Harvard University Press, 1993.

Sadler, Christine. *America's First Ladies.* Macfadden, 1963.

———. *Children in the White House.* G.P. Putnam's Sons, 1967.

Saturday Evening Post. The Presidents: Their Lives, Families and Great Decisions. Curtis, 1993.

Schlesinger, Arthur M., Jr. *The Age of Jackson.* Little, Brown, 1946.

———. *Franklin Roosevelt: Crisis of the Old Order.* Houghton Mifflin, 1957.

Seale, William. *The President's House: A History of the White House.* White House Historical Association/National Geographic Society, 1986.

———. *The White House Papers.* Vol. 2. White House Historical Association/National Geographic Society, 1986.

Shelton, Isabelle. *The White House Today and Yesterday.* Fawcett, 1962.

Shepard, Tazewell, Jr. *John F. Kennedy: Man of the Sea.* William Morrow, 1965.

Sidey, Hugh. *A Very Personal Presidency: Lyndon Johnson in the White House.* Atheneum, 1968.

———. *Portraits of the Presidents: From FDR to Clinton.* Little, Brown, 2000.

Sidey, Hugh; photographs by Fred Ward. *Portrait of a President.* Harper & Row, 1975.

Singleton, Esther. *The Story of the White House*. McClure, 1907.

Slaydon, Ellen. *Washington Wife: Journal, 1897–1919*. Harper & Row, 1962.

Smith, Margaret Bayard, *The First Forty Years of Washington Society*. Charles Scribner's Sons, 1906.

Smith, Marie. *The President's Lady*. Random House, 1964.

———. *Entertaining in the White House*. Acropolis, 1967.

Smith, Marie, and Louise Durbin. *White House Brides*. Acropolis, 1966.

Starling, Colonel Edmund. *Starling of the White House*. Peoples Book Club, 1916.

Strober, Deborah Hart, and Gerald S. Strober. *Richard Nixon: An Oral History of His Presidency*. HarperCollins, 1994.

———. *Ronald Reagan: The Man and His Presidency*. Houghton Mifflin, 1998.

Taft, Mrs. William Howard. *Recollections of Full Years*. Dodd, Mead. 1914.

Thayer, Mary Van Rensselaer. *Jacqueline Bouvier Kennedy*. Doubleday, 1961.

Trollope, Frances. *Domestic Manners of the Americans*. Michael Sadlier, 1927.

Truman, Margaret. *Bess W. Truman*. Macmillan, 1986.

———. *First Ladies*. Random House, 1995.

Truman, Margaret, with Margaret Cousins. *Souvenir: Margaret Truman's Own Story*. McGraw-Hill, 1956.

Tully, Grace. *F.D.R.: My Boss*. Charles Scribner's Sons, 1949.

Van Natta, Don, Jr. *First Off the Tee*. Public Affairs, 2003.

Vidal, Gore. *Lincoln: A Novel*. Random House, 1964.

Wallner, Peter A. *Franklin Pierce: New Hampshire's Favorite Son*. Plaidswede, 2004.

Walton, William. *The Evidence of Washington*. Harper & Row, 1966.

Warner, Judith. *Hillary Clinton: The Inside Story*. Signet, 1993.

Weidenfeld, Sheila Rabb. *First Lady's Lady: With the Fords in the White House*. G.P. Putnam's Sons, 1979.

West, J. B., with Mary Lynn Kotz. *Upstairs at the White House: My Life with the First Ladies*. Coward, McCann & Geoghegan, 1973.

Wharton, Anne Hollingsworth. *Social Life in the Early Republic*. J.B. Lippincott, 1902.

Wilson, Edith Bolling. *My Memoir*. Bobbs-Merrill, 1936.

Wilson, Dorothy Clarke. *Lady Washington: The Story of America's First First Lady*. Doubleday, 1984.

Wister, Owen. *Roosevelt: The Story of a Friendship 1888–1919*. Macmillan, 1930.

Other Selected Sources

Jefferson Rarities Catalog. *Item First Ladies Listing,* 1992, pp. 117–33.

"The Kennedy White House." *White House History.* No. 13 (Summer 2003).

MacDonald, Donald J. "President Truman's Yacht." *Naval History,* Winter 1990.

National First Ladies' Library Bulletin. Vol. 4, no. 1 (Summer 2001).

Pinsker, Matthew. "The Lincolns' Summer Place." *Preservation.* Sept.–Oct. 2003, pp. 55–57.

Salem College Magazine. Summer 2003.

Smithsonian Institution. *The Centennial Post.* Smithsonian Institution and *Washington Post.* Reprint, 1986.

Time. Multiple issues.

Varon, Elizabeth. "Southern Lady, Yankee Spy." *Washington Post,* 2004.

Walker, Mary. Greenwich (Conn.) Library Oral History Project, 1991.

Washington History. Vol. 12, no. 1 (Spring–Summer 2000).

"The White House," *Life* bicentennial issue. Oct. 30, 1992, and other issues.

White House Historical Association. Various publications and website: www.whitehousehistory.org.

White House website: www.whitehouse.gov.

Broadcast Sources

The television shows and series most helpful in my research were *American Presidents* (Public Broadcasting Service); *Biography* (Arts and Entertainment Channel); and various programs on and the History Channel. Special credit to Ken Burns for *Thomas Jefferson;* to Philip B., Philip B. III, and Peter W. Kunhardt for *The American President* and *Echoes from the White House;* and NBC for *Today*'s series of interviews with First Ladies.

Acknowledgments

⊱─◦─⊰

This book has been, in cyberspace terms, a virtual journey of more than two hundred years for me, a fascinating journey that led me back into the forgotten recesses of the nineteenth century, to First Ladies and Children who were no more to me than the paternal family name. Along with the delight of perusing the accounts of early journalists and White House employees, I pulled out my old files and recollections of the First Families I observed and wrote about.

Over the years I stored away an accumulation of insights and anecdotes from many sources, along with my own firsthand observations—even friendships—with a number of First Ladies and Daughters. I am grateful to them all, but thank in particular Lady Bird Johnson, Betty Ford, and Rosalynn Carter for sharing their thoughts in interviews with me in the White House and later. First Daughters Lynda Johnson Robb, Luci Baines Johnson, and Susan Ford Bales reminisced with me, and Julie Nixon Eisenhower and Margaret Truman have contributed to my understanding of life under glass.

A series of First Ladies' staff directors—the East Wing generals—provided guidance that reached far beyond mere facts. Among them, Liz Carpenter set the pace, harnessing the entire country as a backdrop for Lady Bird Johnson's environmental interests; Letitia Baldrige understood a reporter's needs, circumventing the recalcitrant Jackie, and later shared inside stories in her lively book; the late Helen Smith gave me glimpses inside the troubled Nixon White House; Sheila Weidenfeld was always helpful to me and encouraged Betty Ford's forthright cooperation with the press. (Other "generals" who followed would surely be on this list, but by then I was reporting from London and New York.)

This book percolated through years of exchanging viewpoints, judgments, gossip, and curbstone psychoanalysis with fellow journalists as we kept a watchful eye on White House families—especially Frances Lewine (then with the Associated Press, now with CNN), with whom I shared unforgettable adventures tracking Lady Bird Johnson and then Pat Nixon in their travels across the world; we still regularly swap stories from those memorable times.

My special thanks go to Melissa August, who plowed through many arcane volumes for particulars I was seeking and checked innumerable nitpicking details. Clare Crawford-Mason, newspaper, magazine and television journalist, has my undying gratitude for her trove of early tomes portraying private life in the White House; the jewel among them was Ben: Perley Poore's nineteenth-century tour de force.

I found wonderful material in the White House Historical Association's publications and in the Smithsonian Institution's splendid specialized volumes and its new traveling exhibit, First Ladies: Political Role and Public Image. The staff of the Library of Congress gave me prodigious help in searching through its vast collection of old photographs and guided me in the manuscripts room. At the White House, Curator William Allman obligingly dug out the precise details I sought.

The presidential libraries are an ever-growing national treasure, and I made use of all of them—the nation is fortunate to have these unmatched resources protecting and sharing the documents and memorabilia of our history. I must single out Allen Rice of the Nixon Project at the National Archives, Margaret Harman at the Lyndon Baines Johnson Library, and Jessica Sims at the John F. Kennedy Library for their generous assistance in tracking down specific photographs for me.

I am grateful to supportive friends Mary Cronin and Enid Nemy, both sympathetic journalists, and lifelong chum June Arey for their interest in this project, bolstering me when I flagged, and to my agent, Todd Shuster, and the many others who put up with my enthusiasm.

There is a back story to this book and my happy life as a journalist. At a time when women reporters were usually confined to "women's news," I was blessed with editors who gave me the chance to do it all. Most particularly Alicia Patterson, founder and editor *Newsday*, who removed any lingering barriers, and my two bureau chiefs at *Time* magazine, John L. Steele and Hugh Sidey, who opened up the world to me.

And not least is my deep appreciation to Henry Ferris, my editor at William Morrow, for his enthusiasm for this book from the outset and for his elegant touch with his blue pen, a most desirable combination.

Index